ANSWERS
for
ARISTOTLE

Also by Massimo Pigliucci

Nonsense on Stilts: How to Tell Science From Bunk
Making Sense of Evolution (with Jonathan Kaplan)
Denying Evolution

ANSWERS for ARISTOTLE

How Science and Philosophy
Can Lead Us to a
More Meaningful Life

Massimo Pigliucci

BASIC BOOKS
A Member of the Perseus Books Group
New York

Copyright © 2012 by Massimo Pigliucci
Published by Basic Books,
A Member of the Perseus Books Group

All rights reserved. Printed in the United States of America. No part of this book may be reproduced in any manner whatsoever without written permission except in the case of brief quotations embodied in critical articles and reviews. For information, address Basic Books, 250 West 57th Street, 15th Floor, New York, NY 10107.

Books published by Basic Books are available at special discounts for bulk purchases in the United States by corporations, institutions, and other organizations. For more information, please contact the Special Markets Department at the Perseus Books Group, 2300 Chestnut Street, Suite 200, Philadelphia, PA 19103, or call (800) 810-4145, ext. 5000, or email special.markets@perseusbooks.com.

Designed by Linda Mark
Text set in Palatino 10.5

Library of Congress Cataloging-in-Publication Data
Pigliucci, Massimo, 1964–
  Answers for Aristotle : how science and philosophy can lead us to a more meaningful life / Massimo Pigliucci.
    p. cm.
  Includes bibliographical references and index.
  ISBN 978-0-465-02138-3 (hardcover : alk. paper)—
  ISBN 978-0-465-03280-8 (e-book) 1. Life. 2. Ethics. 3. Conduct of life. 4. Science. 5. Philosophy. 6. Aristotle. I. Title.
  BD435.P525 2012
  100—dc23
                                              2012013186

10 9 8 7 6 5 4 3 2 1

CONTENTS

1 Sci-Phi and the Meaning of Life   1

PART I: HOW DO WE TELL RIGHT FROM WRONG?

2 Trolley Dilemmas and How We Make Moral Decisions   21
3 Your Brain on Morality   31
4 The Evolution of Morality   45
5 A Handy-Dandy Menu for Building Your Own Moral Theory   59

PART II: HOW DO WE KNOW WHAT WE THINK WE KNOW?

6 The Not So Rational Animal   77
7 Intuition Versus Rationality, and How to Become Really Good at What You Do   91
8 The Limits of Science   109

PART III: WHO AM I?

9 The (Limited) Power of the Will   127
10 Who's in Charge Anyway? The Zombie Inside You   143

PART IV: LOVE AND FRIENDSHIP

11 The Hormones of Love 157
12 Friendship and the Meaning of Life 173

PART V: THE (POLITICAL) ANIMAL INSIDE YOU

13 Right, Left, Up, Down: On Politics 187
14 Our Innate Sense of Fairness 203
15 On Justice 213

PART VI: WHAT ABOUT GOD?

16 Your Brain on God 231
17 The Evolution of Religion 243
18 Euthyphro's Dilemma: Morality as a Human Problem 263

Conclusion: Human Nature and the Meaning of Life 273

Acknowledgments 289
Digging Deeper 291
Index 307

CHAPTER 1

SCI-PHI AND THE MEANING OF LIFE

"Everything must have a purpose?" asked God.
"Certainly," said man.
"Then I leave it to you to think of one for all this," said God. And He went away.

—Kurt Vonnegut, *Cat's Cradle*

When I was a child I was chubby. This was a cause of major distress, between the taunting of some of my friends and the obvious disapproval of my parents, who kept bringing me to one diet doctor after another. Mind you, by the standards of today's obesity epidemic my condition would have been barely noticeable, and it is quite amusing that nowadays I have to order my clothes online because it is difficult to find clothes in stores that are small enough to fit me. Still, the problem significantly affected my quality of life from the time I can remember being conscious of it (when I was in elementary school) to my early thirties, when I decided that I ought to do something about it, my way.

I began by doing the obvious things: I read about the problem, signed up for a gym membership, and started forming an aesthetic and healthy relationship with food, as opposed to seeing it as a source of consolation for whatever might be going wrong in my life at any particular moment. It took effort, and to some extent it still does, but my quality of life—both physical and psychological—significantly improved in the span of mere months. Without knowing it yet, I was practicing what in this book I call "sci-phi"—short for the wisdom (and practical advice!) that comes from contemplating the world and our lives using the two most powerful approaches to knowledge that human beings have devised so far: philosophy and science.

The basic idea is that there are some things that ought to matter, whatever problem we experience in life: the facts that are pertinent to said problem; the values that guide us as we evaluate those facts; the nature of the problem itself; any possible solutions to it; and the meaningfulness to us of those facts and values and their relevance to the quality of our life. Since science is uniquely well suited to deal with factual knowledge and philosophy deals with (among other things) values, sci-phi seems like a promising way to approach the perennial questions concerning how we construct the meaning of our existence.

So, for example, let's go back to the problem of diet and weight gain. As you might imagine, science has quite a bit to say about this subject, and yet the public is hardly aware of most of this knowledge, which is drowned out by shouts about yet another miracle diet, another miracle pill, or another superficially easy solution to the problem. For instance, Gina Kolata, in her 2008 *Rethinking Thin: The New Science of Weight Loss*, describes a landmark study by Rockefeller University researcher Jules Hirsch, who subjected a group of obese people to a monitored diet while they lived for a grueling eight months at the Rockefeller hospital. Hirsch started out knowing that obese people have much larger fat cells than people of normal weight, and he wondered what would happen to those cells after a rigorous diet: Would they degenerate? Would they get smaller? The results were clear: the fat cells of formerly obese patients had shrunk considerably by the end of the experiment, reaching standard size. Now that his subjects were back to normal, Hirsch thought, not only in overall weight but even on the cellular level, surely they would be able to stay thin, their problem solved. (This was 1959, much earlier than the current obesity epidemic.)

But things did not quite work out as Hirsch predicted: within a few months, all of his formerly obese subjects had gone back to approximately their original weight, despite the fact that they *wanted* to maintain their newly achieved thinness. Since this was science, and the patients were studied from a variety of perspectives under controlled conditions, the researchers were able to figure out what had happened. The metabolism of an obese person is actually normal, meaning that it is calibrated by the body to maintain the status quo. When the researchers put their obese subjects on a severe diet, they of course lost weight (simply as a result of basic laws of physics), but their metabolism slowed down enormously compared to that of a naturally thin person. In other words, their bodies interpreted the new regime as one of starvation, and a basic survival mechanism (keep the metabolism down, consume less) kicked in. Once the dietary restrictions were lifted, the subjects' metabolism levels went back to normal, they felt an uncontrollable hunger, and they ate their way back to obesity.

Later studies not only confirmed these findings but uncovered the symmetrical truth that thin people who attempt to gain weight (sometimes by eating as much as an astounding 10,000 calories a day!) cause their metabolism to accelerate dramatically. As soon as they stop overeating their bodies burn all the extra fat, and they are back where they started. (Lucky bastards, one might add.) This kind of research, as well as studies on the genetic inheritance of body mass index (a function of your weight, and your height), all point to the conclusion that most of us have a "set range" in terms of both our metabolism and our weight and that our bodies become increasingly resistant to attempts to move much above or below our natural range. This most certainly does not mean that the environment has nothing to do with our weight, or that we cannot do anything about it. But it does mean that not only are there limits to what we can do, but there is a cost to be paid for doing it in terms of the precious psychological resource of willpower (more on that in Chapter 9). If people were more widely aware of this sort of research, they would have more realistic expectations about how to handle their weight problem, they would devise smarter ways to deal with it, and they would stop pursuing the chimera of a "silver bullet" that will quickly make them happy. Then again, the huge cottage industry that has flourished by exploiting people's weaknesses about food would probably collapse overnight, with presumably disastrous consequences for all those who keep making money out of the craze for easy diets.

So much for the science. Where does the philosophy come in? All the facts about the inheritance of human weight, rates of metabolism, the size of fat cells, and so on are of only academic interest until they affect the lives of real people. But why do these facts affect our lives? One answer comes, again, from science: there are negative health consequences to being seriously overweight or obese. People who suffer from severe weight problems are more prone to diabetes and heart conditions, their life expectancy is significantly shorter, and of course the quality of that life is much diminished. Moreover, there are practical consequences for society, which devotes significant financial and other resources to treating conditions related to obesity.

But that is, of course, only part of the story. I was never obese, not even close, and yet the problem of being overweight has been with me my entire life. And I am most certainly not an isolated case: a multimillion-dollar industry of diets and exercise machines exploits the obsession with weight shared by millions of Americans. What is going on is that we make judgments concerning the issue of weight, judgments that range from aesthetic to moral in nature, and both aesthetics and ethics are branches of philosophy, not science.

If we feel ugly because we are overweight, it is because—probably unconsciously—we deploy a certain aesthetic theory of what an attractive human being should look like. This theory is informed, of course, by the culture in which we live (the concept of physical beauty has demonstrably changed over space and time throughout human history) and perhaps to some extent even by basic biological instincts. (There is evidence, for instance, that we prefer symmetrical features in a potential partner because such features are a reliable indicator of healthy genes that could be transmitted to our progeny.) Similarly, if we blame our excess fat on our lack of self-control, we are making a moral judgment about how we *ought* to live and behave, what we should strive for in life, and how much of our resources (both mental and financial) we should invest in achieving certain aesthetic standards. We are doing philosophy without realizing it, and there is a distinct possibility that bad philosophizing may make our lives more miserable than they perhaps should be.

This idea that philosophy and science can be combined to give us the best possible knowledge about the world and how to act within it is an old one, encapsulated by the classic concept of *scientia*, a Latin word that means "knowledge" in the broader sense, encompassing both the sciences and the humanities. In German, a similar term is *wissenschaften*, which also refers to more than just the English "science." Arguably the first philosopher in the Western tradition to take the concept of scientia (or what I call sci-phi) seriously was Aristotle (384–322 BCE), a fellow whom the reader will see popping up throughout this book (which explains the title of the same). What is important about Aristotle and some of his fellow ancient Greek philosophers is not the content of their science, which has been dramatically outpaced by the developments of the past twenty-four centuries, and not even some of their specific philosophical positions, which are also no longer tenable in light of subsequent discussions. Rather, what is crucial to the idea of this book is the ancient Greeks' fundamental concept that life is a project and the most important thing for us to do is to ask ourselves how we are to pursue it. In a sense, then, Aristotle was among the first to approach the big questions in both a philosophical and a scientific manner, and we are now beginning to have some good (if still provisional) answers to those questions.

For Aristotle this project meant engaging in a quest for *eudaimonia*, a Greek word that literally means "having a good demon" and that is often translated as "happiness," though it should more properly be understood as "flourishing." Eudaimonia is achieved by engaging in virtuous behavior—that is, doing the right things for the right reasons—throughout one's existence. Since life thus conceived is a project, a full assessment of a life's worth is actually not possible until we reach the end, a notion that still has a powerful intuitive appeal for us moderns. For instance, the lifelong reputation of someone who led a good life up to a certain point but then engaged in unethical behavior is diminished or crippled, and vice versa: we consider praiseworthy someone who began by faltering but then regained a high ethical ground.

Aristotle, a good psychologist, realized that it is often difficult to align our rational assessment of what we ought to do with our emotional inclinations toward doing what comes easily to us or is pleasurable. Going back to our example of physical well-being, we all know that it is good for our long-term bodily (and psychological) fitness to eat moderately and exercise regularly. Yet our penchant for immediate rewards pushes us to overindulge our appetites and makes us lazy when it's time to get on the treadmill. Aristotle never saw a fast-food joint, but he knew about human fallibility. Indeed, he thought that the major obstacle to increasing our eudaimonia was something the Greeks called *akrasia*, translatable as "weakness of the will." In a sense, to be virtuous means to rise above one's weaknesses to do the right thing, both for ourselves and for others. That is the way toward human flourishing.

It should be clear, then, that eudaimonia is not "happiness" in the generic sense of a positive emotion, but that it is value-laden—that is, it is an intrinsically moral concept. The ancient Greeks would have argued, for instance, that a life devoted to the exclusive pursuit of physical pleasures, or wealth, or power, is not eudaimonic, regardless of how "happy" the person seeking such pleasures, wealth, or power might feel. In an important sense, these pursuits would all be distractions from eudaimonia, because the individual in question has done nothing to improve himself or to affect the world in a positive manner. Eudaimonia should not be confused with the Christian virtue of asceticism, or with Buddhist detachment: the Greeks from Aristotle to Epicurus (341–270 BCE) thought that physical pleasures like good food, love and friendship, and even a good dose of luck were all necessary ingredients of a eudaimonic life. They also thought, however, that we are best able to pursue the eudaimonic life by taking the time to reflect on it.

All of this notwithstanding, to most people these days philosophy seems like a quaint activity best left to a bunch of old white men with a conspicuous degree of social awkwardness. This is the twenty-first century: if science tells us that a certain weight range is good for our health and another is likely to trigger disease, shouldn't a rational human being simply follow the doctor's orders, philosophical disquisitions about aesthetics, ethics, and the meaning of life be damned? I don't think the answer to this question is quite so simple, because of a crucial and much underappreciated distinction between facts and values. To derive the latter from the former is a type of logical error known as the "naturalistic fallacy" (because one is attempting to equate what is natural with what is good). One of the first to discuss the naturalistic fallacy (though he didn't use that term) was the eighteenth-century Scottish philosopher David Hume (1711–1776) in his aptly titled *A Treatise of Human Nature*. Hume noticed that some people who wrote on a variety of factual issues (what is/what is not) eventually, seamlessly, and without explanation switched to an altogether different kind of discourse concerned with ethical imperatives (what ought to be/what ought not to be). Hume is not saying that there is no connection between facts and values, but he points out that a person invoking such a connection should explicitly justify it.

Hume's conception of the naturalistic fallacy informs this book and its central idea that the conjunction of science and philosophy has much to offer in making the lives of reasonable human beings significantly better. Taking the naturalistic fallacy seriously, we acknowledge that science (dealing with matters of fact) is not enough; we also need philosophy (dealing with matters of value). But our philosophy can and should be informed by the best science available, and vice versa: our quest for scientific knowledge should be guided by our values. Our aesthetic judgment may want our bodies to be close to a particular weight range, and our moral judgment may fault us for not quite getting there. But a serious understanding of the biology of human metabolism will help us cut ourselves some slack and reach a compromise between what we would like to do and what biological reality allows us to do. Here science helps us revise our philosophical intuitions. That science should in turn be guided by our philosophical choices is also clear upon a moment's reflection: why does so much money go into research on human weight loss and gain? Because according to our aesthetic and ethical values, that sort of investment is justified, presumably at the expense of other possible kinds of medical research, considering that our societal resources are not unlimited. Philosophy, then, guides the general direction in which science (and science funding) goes.

Of course, a clear distinction has been made between science and philosophy only rather recently in human history; not so long ago, in the seventeenth and eighteenth centuries, people like Galileo Galilei (1564–1642) and Isaac Newton

(1642–1727)—who today would be considered scientists—saw themselves as "natural philosophers." If science and philosophy were once one and the same, how do we account for their peculiar evolution so that nowadays they are treated as distinct enough to often be housed in different colleges on university campuses? And what can I possibly mean by "sci-phi," the attempt to bring them back together in the service of human flourishing?

The evolution of science is perhaps easier to understand, if for no other reason than that more people are familiar with its products, from its discoveries about the nature of the world to the technological and medical applications of its principles. This familiarity notwithstanding, there are many misconceptions about science, which I tried to clear up in *Nonsense on Stilts: How to Tell Science from Bunk*. To begin with, there is no such thing as the "scientific method." Science is a somewhat methodical enterprise, but every practicing scientist has one basic guiding principle: whatever works. Scientists are by nature pragmatic, and they will approach a problem from a variety of points of view, deploying an array of methods of investigation, until they reach a satisfactory answer to their questions.

One of the uncanny things about science is that, in often reaching improbable conclusions about how the world works, it repeatedly defies common sense and provokes widespread rejection of its findings. We now think, for instance, that quantum effects (in particular, the Pauli principle) are responsible for the fact that solid objects occupy space; we know that our planet is a speck of dust at the periphery of a giant galaxy, itself just one of many billions that populate the universe; and we have excellent reasons to believe—contrary to a surprisingly still popular opinion—that human beings are very close relatives of chimpanzees and gorillas. Science proceeds in a way similar to Sherlock Holmes's explanation to Dr. Watson in "The Adventure of the Beryl Coronet": "It is an old maxim of mine that when you have excluded the impossible, whatever remains, however improbable, must be the truth."

Another commonly misunderstood characteristic of science is that it is not in the business of delivering permanent truths, but offers instead provisional conclusions that have a certain probability of being true. So, for instance, when I said in the previous paragraph that we know that our planet orbits around an average star in the suburbs of the Milky Way galaxy, what I meant was that this is the best supported conclusion arising from a myriad of astronomical observations and from our theoretical understanding of planets, stars, and galaxies. It is certainly possible that either the observations or our theories (or even both!) may turn out to be wrong or deeply flawed in some manner, and that future generations of astronomers will look upon us with the same condescending smile that we reserve for Ptolemy's idea that the earth was the center of the universe and the rest of the cosmos went around it, moved by invisible celestial spheres.

The tentativeness of scientific conclusions is a source of continuous inspiration to scientists, but also of perennial frustration and misunderstandings for policymakers and the general public, all of whom would much rather be told "the Truth" by scientists and be done with it—especially after paying millions of dollars to finance scientific research. And yet there is a profound lesson in humility to bring home here. It is rather ironic that science is often portrayed as the ultimate refuge of the arrogant, but that scientists themselves keep trying to explain to us the limits inherent in the human quest for knowledge about the world (discussed in Chapter 8).

What, then, constitutes science as a distinct field, separate from philosophy, literary criticism, or whatever else? Although a precise definition of science is probably impossible because of the nature of the beast itself, I would say that science is a form of inquiry into the natural world characterized by the continuous refinement of theories that are in one way or another empirically verifiable. It is this unique blend of theorizing and empirical investigation that is at the core of the scientific enterprise. As the philosopher Immanuel Kant (1724–1804) famously put it, "Experience without theory is blind, but theory without experience is mere intellectual play." If science were just about facts, it would be the same as stamp collecting, but if scientific theories were not continuously checked against empirically verifiable facts (through either experiment or observation), they would quickly degenerate into pseudoscience, the kind of thing we see with astrology, parapsychology, or creationism.

What about philosophy? If science is difficult to define precisely, the task is pretty much hopeless in the case of philosophy, a much older discipline that has evolved along very distinct and sometimes contrasting lines. Broadly speaking, philosophy is traditionally divided into a number of branches that deal with questions concerning the nature of reality (metaphysics), our access to that reality (epistemology), what we ought or ought not to do (ethics), how we should reason (logic), and what is beauty (aesthetics). More recently, the emergence of a spate of new fields—typically labeled "philosophy of science," "philosophy of mind," "philosophy of religion," and so on—has advanced concerns with the philosophical aspects of other disciplines.

One of the most influential philosophers of the twentieth century, Ludwig Wittgenstein (1889–1951), said, "Philosophy is a battle against the bewitchment of our intelligence by means of language," and that is certainly a good way of looking at what philosophers do. What Wittgenstein meant was that human language is by its nature imprecise and prone to confusion (providing an ever-renewable source of material for comedy, which in turn has seriously been compared to philosophizing) and that we therefore ought to be perennially on guard against being misled by how we use words. Then again, it can also be argued that sophisticated thinking about the world simply cannot be achieved without language, so we seem to be in a bind. The problem is no different, in principle, from the one that scientists encounter in their line of work: every tool they use is by necessity limited and flawed in some respect, and yet they need to use those tools to move forward with their investigations. The difference is that philosophy, in this way of looking at things, deals with the power and limits of the ultimate human tool: language itself. (Incidentally, this is not to say that philosophy reduces to linguistics; to make things even more confusing, you might not be surprised to learn that there is also a philosophy of language.)

Of course, if you asked one hundred philosophers what philosophy is, you would probably receive the proverbial hundred (and one) different answers. And yet, thinking of philosophy as the discipline that deals with the rational use of language seems to me the easiest way to understand why philosophy also is the broadest possible discipline: because it deals with our most basic tool for knowing and communicating things, it in some sense encompasses all of human knowledge. There are plenty of other conceptions of philosophy, but my take on it is that ultimately philosophy is founded on the construction (and deconstruction) of reasoned arguments. Typically, a philosopher will pose herself a set of questions, examine what is known about them, and explicitly reason her way to a particular conclusion. Other philosophers will then examine and pick apart her reasoning and see how it withstands critical scrutiny, probably presenting arguments of their own in favor of a different conclusion. And so on. There are traditions that are commonly counted as philosophy that do not follow this modus operandi—for instance, the so-called Eastern philosophies, and also part of what is today referred to as "continental" philosophy (because it originated with the work of some eighteenth- and nineteenth-century philosophers from continental Europe, such as Immanuel Kant and Friedrich Nietzsche [1844–1900]). What I consider philosophy in this book is the sort of intellectual activity that started out in ancient Greece before Socrates (469–399 BCE) and whose Greek root charmingly translates to "love of wisdom."

Just as with science, there are several common misconceptions about philosophy, and one definitely needs to be put to rest at the outset: philosophy, like science, does make progress, even though its progress has to be measured differently from scientific progress. Science, roughly speaking, can be said to make progress in proportion to how its understanding of the world matches the way the world actually is. (This idea is a bit simplistic, as any good philosopher of science will tell you, but we will take it as a good enough approximation for our purposes.) For instance, the Copernican theory of the solar system is better than the Ptolemaic one because, as a matter of fact, it really is the sun and not the earth that is at the center of the system. And the Keplerian theory that succeeded the Copernican one is even better, because it comes even closer to reality: Kepler (1571–1630) realized that the planets follow elliptical orbits, not circular ones, as Copernicus (1473–1543) thought, and that the sun is not quite at the center of the whole thing but rather occupies one of the foci of those ellipses.

Analogously, philosophy also makes progress when it understands better and better the meaning and implications of human concepts and how they relate to the world. For instance, philosophers have produced several theories of human morality, exploring a variety of logical possibilities (discussed in Chapter 5). As I have already hinted, Aristotle said that morality is about what makes human beings flourish (so-called *virtue* ethics); Jeremy Bentham (1748–1832) and John Stuart Mill (1806–1873) in the eighteenth and nineteenth centuries advanced the idea of *utilitarianism*, according to which what is ethical is whatever increases the happiness of the greatest number of people; and Immanuel Kant, also in the eighteenth century, articulated a view of morality as a set of rules based on certain duties we ought to have with regard to other human beings (a rule-based, or *deontological*, ethics).

Philosophers have worked out the implications of these and other systems, criticized them, and proposed refinements and alternatives. No philosopher today would be so naive as to espouse any of these ideas in anything like their original form, because discussions in the field have led to more sophisticated versions of them, and indeed, the debate is still moving forward from this new level of understanding.

Regardless of how we think about philosophy as a discipline, its relationship with science will take some interesting turns in this book. As I mentioned earlier, science started out as a branch of philosophy known as natural philosophy, and it stayed that way until the seventeenth century or thereabouts. With the advent of modern academic specialization, however, not to mention the explosion of discoveries in science, the two fields have become largely separated. There are some interesting fields that are still at the borderline today and that show us how areas of philosophy can turn into corresponding areas of science. One of them is philosophy of mind, which concerns the nature of consciousness. Until very recently, this was an exclusively philosophical enterprise, but more and more neuroscientists have tackled the question because of the availability of new experimental techniques; functional magnetic resonance imaging (fMRI), for instance, now permits a researcher to note which parts of the brain are active when a subject is immersed in particular mental tasks, such as reading, or thinking about ice cream.

Nowadays academic conferences and journals dedicated to the study of consciousness are populated by both philosophers and scientists, and my guess is that the field will gradually come to be dominated by scientists, as has already happened in, say, psychology (which, before William James [1842–1910], was also a branch of philosophy). Such an evolution does not reflect the relative value of philosophy and science, but rather results from the fact that the two approaches are complementary: when a problem is vaguely defined and empirically unassailable, it is the task of philosophers to clarify matters and prepare the conceptual ground for the time when science has developed the appropriate experimental tools to investigate it.

But this sort of transition is not the only possible path. There will probably always be questions for which Hume's naturalistic fallacy precludes a transition between fields. Indeed, these are the cases that interest us the most in this book. Morality will figure in much of what we discuss here, for the very good reason that the moral sphere makes up a lot of what goes into making our existence meaningful. The naturalistic fallacy prevents us from simply accepting scientific answers (is/is not) to moral questions (ought/ought not), though our philosophical discussions of meaning and values should most certainly be informed by the best available scientific understanding of the relevant issues.

As we will see, there are plenty of other areas where the combined insights of the best science and the best philosophy make it easier for us to consider our problems from a more informed, more rational position. A conceptual map of the territory ahead in this book would range from how to tell right from wrong to what counts as knowledge and why; from considerations of who we are to discussions of love and friendship; from an analysis of justice and politics to the ever-present issue of gods and what they might contribute—if anything—to the meaningfulness of our existence. In all of this, the practice of what I call sci-phi makes one crucial assumption: that you are interested in using reason and evidence to guide your life and make it better. If you'd rather be led by mysticism, superstition, or "other ways of knowledge"—whatever they may be and however they may work—this is not the book for you. But if you agree that the most precious organ a human being possesses is her brain and that we owe it to ourselves to make the best use of it that we can, then by all means turn the page and let's get started.

PART I

HOW DO WE TELL RIGHT FROM WRONG?

CHAPTER 2

TROLLEY DILEMMAS AND HOW WE MAKE MORAL DECISIONS

> Moral excellence comes about as a result of habit. We become just by doing just acts, temperate by doing temperate acts, brave by doing brave acts.
>
> —Aristotle

IMAGINE THAT YOU ARE AT THE HELM OF A TROLLEY, GINGERLY carrying passengers from one stop to another along the city's streets. Suddenly you see five people standing close, right in front of you, about to be hit by the trolley! You slam on the brakes, and they don't work! You shout to them to get the hell out of your way, but they don't see or hear you! In desperation, you realize that there is only one option available to you: the trolley tracks are about to split, and if you pull a lever you will change course and save the five people. In so doing, however, you will inevitably end up hitting, and probably killing, an innocent bystander. Would you pull the lever?

Most people who are surveyed answer affirmatively, if reluctantly, across cultures. This is one version of one of ethical philosophers' favorite thought experiments, the trolley dilemma. The idea of course is to see what people's moral intuitions are when faced with difficult ethical quandaries. The results would suggest that most people adopt what philosophers call a utilitarian, or *consequentialist*, form of moral decision-making: saving five people is the right thing to do, even though another person will be killed in the process. It's an example of what Jeremy Bentham—the originator of utilitarianism—would call "moral calculus."

Almost every time I explain the trolley dilemma someone inevitably begins to rattle out the obvious objections: But what if the five people are all Nazis and the one you kill is your mother? Could you not alert the people to move? Is there no other option available? Or some variant thereof. But the point of the experiment is precisely that there are no other options, and that we do not know anything about the people involved. This kind of experiment allows us to examine people's moral intuitions, other things being equal. Incidentally, the scenario may appear far-fetched, but it isn't. There are plenty of situations in real life, from medical emergency rooms to police or military actions, where people are suddenly confronted with the need to exercise some sort of moral calculus quickly, with few available options, and on the basis of very little information. Philosophy sometimes is a question of life and death.

The trolley exercise does not stop there. A variant of the trolley situation reveals something even more interesting about how we think about morality. Imagine now that instead of driving the trolley you are walking on a bridge, below which you see the trolley, on its tracks, about to hit the five bystanders. Now your only course of action, the only way to save the five people, is to quickly grab a large person standing near you and throw him off the bridge, thereby blocking the advancing trolley. Would you do it? (Again, no other options are available; you cannot sacrifice yourself, perhaps because your body mass is too small to stop the advancing car.) It turns out, somewhat surprisingly to philosophers, that in this version of the dilemma most people recoil from sacrificing the one for the many. This is unexpected because clearly the new course of action is not consequentialist at all. Rather, it seems to fit with some other general way of thinking about morality, perhaps a type of *deontology*, or rule-based ethics, similar to the ones adopted by many religions. (The Ten Commandments are the obvious example.) Maybe the rule here is "Thou shall not kill an innocent person," or the Kantian imperative not to use other people as means to ends. Yet that can't be the whole story, because people who adopt a deontological morality in the bridge version of the dilemma are directly contradicting their obviously consequentialist approach in the lever version. We return to the problem created by contradictory ethical doctrines in Chapter 5.

At any rate, here is where cognitive scientists enter the picture. A group of researchers led by Michael Koenigs of the University of Iowa and Antonio Damasio of the University of Southern California in Los Angeles performed an interesting neurobiological experiment using the trolley problem. They compared normal subjects with people who had suffered a specific kind of neurological damage in the ventromedial prefrontal cortex of the brain, an area known to affect emotional reactions. After presenting both sets of subjects with both versions of the trolley dilemma, Koenigs and his colleagues discovered something very interesting about how the brain works when it is engaged in moral decision-making. There was no difference between normal subjects and neurologically damaged patients in their responses to the lever version of the dilemma: most subjects in both groups agreed that it was acceptable to pull the lever, thereby trading five lives for one. However, when faced with the bridge version of the problem, twice as many brain-damaged patients said that the utilitarian trade-off was still acceptable (that is, that it was okay to throw the big guy off the bridge) compared to the controls. What gives?

Joshua Greene, a cognitive scientist at Harvard, thinks he knows what's going on. His group has shown that different areas of the brain are activated when someone is considering a personal versus an impersonal ethical problem—just like the difference between the bridge and lever versions of the trolley dilemma. Predictably, the lever situation elicits a strong response from areas like the medial frontal gyrus, which is known to be associated with emotions, while the bridge situation stimulates those sections of the brain known to be involved in problem-solving and abstract reasoning. The difference between the brain-damaged and normal subjects in the Koenigs study, then, is explained by the fact that their emotional circuits were impaired, so to speak.

So, does the available science favor consequentialism or deontological ethics? A philosopher would say that this is a strange question, since neurobiology can tell us how people think, but not how they *should* think. Indeed, a naive scientist could make the claim that the neurobiological evidence favors a consequentialist ethical philosophy over a deontological one based on the observation that when people use their reasoning faculties—and isn't that the logical thing to do?—they "go consequentialist." Then again, a neuroscientist with Kantian inclinations could equally reasonably point out that it is the people with incapacitated emotions—people whose brain isn't working the way it is supposed to—who favor the consequentialist solution in the bridge version of the dilemma. You can see how simple facts, however interesting they may be, are just not enough to decide what's the right thing to do.

Jonathan Haidt is a social psychologist who has made some intriguing observations about the question of human moral judgment. He has proposed the "mere rationalization hypothesis," which essentially states that a lot of our moral decisions arise from evolutionarily ingrained instincts or emotions and are not ethical at all. Haidt refers to a study he conducted in which he exposed subjects to actions that caused no harm and yet were likely to provoke a strong emotional reaction. For instance, he looked at how people responded to the idea of cleaning the toilet with one's national flag. Predictably, most people recoiled from the action, and when asked to elaborate they produced explanations that used moral terms to condemn such a use of the flag. But, Haidt argues, since these actions do not actually harm anyone, in what philosophically coherent sense can they be considered immoral? Instead, he suggests, this is one example of people *rationalizing* their evolutionarily or culturally engrained emotions and dressing them up as moral when in fact they are arbitrary. According to Haidt, we should learn to distinguish valid moral judgments from those caused by our evolutionary or cultural background and make an effort to discard the latter in favor of the former.

But then the obvious question becomes: How do we tell the difference between spurious and valid moral explanations? Why doesn't the mere rationalization hypothesis hold all the way down, so to speak, providing scientific backing for the "anything goes" idea of moral relativism? Philosopher William Fitzpatrick points out that in some cases we can clearly distinguish between evolutionary and ethical considerations, as when people make decisions that seem to be guided by moral reasoning that flies straight in the face of their evolutionary instincts. For instance, we may decide not to have more than two children because we are concerned about world population (thus violating the Darwinian imperative to reproduce as much as possible); or we may give up part of our time to volunteer for a humanitarian organization; or we may send a check to a charitable organization so that a child on the other side of the world will have a chance at survival, health care, or education; or, at the extreme, we may even sacrifice our own life for a cause we deem worthy (which amounts to nothing less than evolutionary suicide). None of these decisions make sense from a purely biological standpoint, which would have us focus our efforts on two and only two things: survival and reproduction (and the first imperative is important, from the point of view of natural selection, only if it leads to the second one).

The widespread existence of human behaviors like the ones just mentioned (and many others, of course) is a real problem for any strong evolutionary theory of morality. Still, Fitzpatrick points out, such behaviors do not mean that evolution has no bearing on why we are moral animals. He articulates what he calls a "modest evolutionary explanatory thesis," according to which our evolutionary history tells us something about why we have the tendency and capacity for moral thinking, as well as why our moral thinking is accompanied by certain emotions. We will examine what evolutionary biology has to say about morality in more depth in Chapter 4. For now, we are still left with a serious problem. Consider the following moral judgments (MJs) (again, from Fitzpatrick's work):

MJ1: Interracial marriage is wrong.
MJ2: Homosexuality is wrong.

Until recently, both MJ1 and MJ2 were considered true in Western societies, and both are still considered valid in many non-Western societies. However, most Westerners have moved away from MJ1, and an increasing number of them have also abandoned—or at least seriously questioned—MJ2. But the moral skeptic would obviously say: Doesn't this variety of opinions clearly show that moral judgments are culturally relative? That what is morally "true" in one place or time is not necessarily true in another cultural or temporal context? This is a crucial question. We saw in the last chapter that we cannot derive moral oughts from matters of fact (at least according to Hume). If it now turns out that we have no reason-based approach that leads us to say that something is moral or not, then the relativist might win the field after all, leading to a situation where we have no moral guidance other than our tastes and idiosyncrasies.

This is the territory of so-called *metaethics*—the discipline that examines the rational justifications for adopting any moral system at all (as opposed to ethics, the branch of philosophy that debates the relative merits of different views of morality and how they apply to individual cases). Metaethical issues are notoriously hard to settle, for a reason very similar to why it has proven doggedly difficult to provide rational foundations even for mathematics and logic, the quintessential areas of pure reasoning. Throughout the twentieth century the top logicians in the world embarked on an epic quest to find a tight logical foundation for mathematics (a quest delightfully recounted in the graphic novel *Logicomix* by Apostolos Doxiadis and Christos Papadimitriou). That search for the holy grail of reason ended in defeat when Kurt Gödel (1906–1978) demonstrated (logically!) in 1931 that it just wasn't possible to find such a foundation. Then again, Gödel's so-called *incompleteness theorem* has not induced mathematicians to hang up their pencils and paper and go fishing, so maybe we too can set metaethics aside for another day without having to give up the idea of moral reasoning.

That said, it turns out that philosophers think that neither MJ1 nor MJ2 are valid moral judgments. Moreover, they think that the following moral judgment is a valid one:

MJ3: The unmotivated killing of another human being is wrong.

Why? Fitzpatrick summarizes what a philosopher would say about MJ1 and MJ2 in this way: first, both statements fail to withstand critical reflection; second, the reason some people think that MJ1 and MJ2 are true (even though they are not) is nonmoral in nature.

Let's start with the second criterion: it is easy to attribute an endorsement of MJ1 to racism and an endorsement of MJ2 to homophobia, both of which explanations can be tested independently (that is, we can tell via other means whether a person is racist or homophobic). In the case of MJ3, however, it is hard to think of a nonmoral motive for the judgment.

The first criterion is, of course, trickier, as the type of critical reflection one applies to the alleged moral judgments depends on what type of ethical system (consequentialism, deontology, and so on) one accepts more generally. Still, we could argue that both MJ1 and MJ2 are wrong for a variety of reasons: they discriminate against an arbitrary group of people (members of another race, homosexuals); we would not want to have these sorts of judgments applied to our own decisions in matters of marriage and sexual practices; or such prohibitions infringe on personal liberty in situations where no one is being harmed by individuals' choices. MJ3, by contrast, stands up to such critical evaluation because, if we did allow random killings, we would soon not have a society to speak of, since a society is a group of individuals who band together for reasons that include increased personal safety. (Notice, of course, that the word *unmotivated* in MJ3 is a big caveat: it allows for the moral acceptability of killing someone in, say, self-defense, or for other reasons that need to be specified and analyzed. The point is that such reasons cannot be arbitrary and at the whim of cultural trends.)

To take stock, it would seem that moral judgment is still an area where philosophy dominates, because it is hard to justify the equation of what is natural (as in the result of evolutionary processes, or the brains' way of connecting analytical thinking and emotional reactions) with what is right. This does not mean, of course, that philosophers have an easy time settling ethical disputes or even rationally justifying why we should be ethical to begin with. Still, science does tell us quite a bit about how our brains work when we do exercise ethical judgments, and even about how we acquired this somewhat strange idea that there are "right" and "wrong" things out there. In the next two chapters, we turn to more neurobiology and evolutionary biology to help us make sense of what it means to be a moral animal.

CHAPTER 3

YOUR BRAIN ON MORALITY

> As for morality, well, that's all tied up with the question of consciousness.
>
> —Roger Penrose

IN 1667 ONE THOMAS CORNELL WAS HANGED BECAUSE he had been found guilty of murdering his mother. A little more than two hundred years later, one of his descendants, Lizzy Borden, was controversially acquitted of charges of having killed her father and stepmother. At the onset of the twenty-first century, yet another descendant of the same family, Jim Fallon, is a professor at the University of California at Irvine, where he studies the brains of serial killers. The interesting thing is that until a few years ago Fallon was not aware of his family's, shall we say, interesting history or of how pertinent that history was to his own academic interests. It apparently was after a casual conversation with his mother that he began to look into it, and the more he looked the more worried he got.

To satisfy his own curiosity, he had several members of his family brain-scanned, including himself. You see, Fallon's research shows that serial killers tend to have very little activity in the area of the orbital cortex. This makes sense, because that area of the brain is known to interact with and repress the activity of the amygdala, which—to simplify a bit—is the seat of our strong emotions, particularly fear, but also the spring for our aggressive behavior. No activity in the orbital cortex means that the normal brakes on the amygdala have been lifted, so to speak, making an individual more prone to violence. None of Fallon's close relatives turned out to have the brain signature of a serial killer—but he did!

At this point the biologist began to feel a bit uneasy, but he pressed on with his quest nonetheless. There was a second test he could run that would be pertinent, one that didn't deal directly with the structure of the brain, but rather with the genetic bases of aggression. The monoamine oxidase-A (MAO-A) gene is found in different variants in the human population, just like most genes. It happens, however, that one of these variants is associated with particularly violent behavior and, again, is frequently found among serial killers. That variant, nicknamed "the warrior gene," was absent from the DNA of Fallon's relatives, but as I'm sure you'll be less than surprised to discover, he had it. And yet, Jim Fallon is not a serial killer—he just has an academic interest in the phenomenon. What is going on?

Welcome to the increasingly fascinating field of *neuroethics*, where philosophers and scientists come together to better understand (and perhaps improve) the way human beings reason and act from a moral perspective. This book is about what philosophy and science together can tell us concerning the big questions in life, and if we want to understand these questions in a new light we also need to look under the hood, so to speak. We will employ not only the logical scalpel of philosophy to parse what people mean by the different ideas that guide their lives but also the microscope of science to try to figure out how and why people behave in certain ways. In this chapter, then, we focus on the hows of moral reasoning from a neurobiological perspective. We will turn to the whys—what we can tell about the evolution of morality—in Chapter 4. And we will wrap up this topic by returning to philosophy in Chapter 5, this time armed with a better understanding and ready for better guidance on how to live an ethical life.

Jim Fallon does have an idea about why he is not a serial killer. Despite his family history, being a carrier of the warrior gene, and having the characteristic deadness in his orbital cortex, another element is missing, he believes: the right (or rather, wrong) environment. Fallon had a nice childhood with no traumas and plenty of affection from his family, but if it had been different—had he been abused, for instance—then the perfect neuro-genetic-environmental storm would have been unleashed, he thinks, and he might have been a subject for someone else's studies on serial killers. Perhaps, but we do not know—we can only speculate about such things. At the very least, the strange case of Jim Fallon highlights the fact that particular genetic or neurological signatures are not *sufficient* to trigger a given set of behaviors. They may, however, be sufficiently important factors to be admitted in a court of law.

According to a National Public Radio investigative report in 2010, American courts have allowed evidence about the neurobiology or genetics of violent behavior in about 1,200 cases so far, and it looks like this is just the beginning of a trend. For instance, in 2006 in Tennessee one Bradley Waldroup was accused of killing his wife and a female friend of his wife's during a violent outburst at the end of an altercation. From a forensic point of view, Waldroup's culpability was obvious and the prosecutors asked for the death penalty. But the defense attorney argued that evidence should be admitted to the effect that Waldroup had the very same MAO-A variant, the "warrior gene," that Jim Fallon found himself carrying. The attorney argued that the defendant was prone to snap under pressure and engage in violent acts because of his genes, and in a stunning outcome the jury agreed: Waldroup was convicted of voluntary manslaughter and avoided the death penalty.

From a philosophical perspective, there are two reasonable ways of looking at this case, and they carry us to very different conclusions. On the one hand, it is a well-established principle of modern American law that people with extremely low intelligence should not be sentenced to death, even if they have demonstrably committed a crime for which capital punishment might otherwise be considered. (Remember that the United States is the only Western country where the death penalty is possible to begin with.) The reasoning behind this principle is that, because such people are incapable of the same degree of understanding and decision-making that most of us can muster, the ethical thing to do is to *restrain* them from doing additional harm, but not to punish them for something over which they had little deliberative power. On the other hand, there clearly has to be a limit to how much biological considerations can enter into our system of laws or the concept of justice will simply lose any coherence. If the defense is that "my brain made me do it," or "my genes made me do it," simply consider that pretty much anything we do is affected by our genetic makeup, and certainly our brains get involved in everything we do. You see the dilemma.

Moreover, our moral judgments can be skewed by factors that are not nearly as dramatic as having a silent orbital cortex or a warrior gene. For instance, what if I told you that watching an episode of *Saturday Night Live* affects not just your mood (if you appreciate that sort of comedy) but measurably alters your immediate moral judgment? (You become more of a utilitarian, or consequentialist, if you watch comedy.) Or how about the fact that if I were to ask you about an ethical issue while you were sitting at a dirty desk or smelling an unpleasant odor, you would be more likely to render a severe judgment than if you were at a clean desk or your nostrils were not under assault? Clearly much more than calm and rational deliberation goes into our moral decision-making, and indeed, much of what influences that decision-making flies quite easily below our conscious radar—unless we know it's there and we keep our guard high.

We already encountered some of these additional factors in Chapter 2, when we considered sci-phi research into the trolley dilemmas, and now it is time to return to the ideas of one of the scientists we have already met, Joshua Greene of Harvard. Greene has reviewed much of the literature on the neurobiology of moral decision-making and has come up with what he calls a "dual-process" theory of moral judgment. According to Greene's theory, we change the type of moral judgment we employ—going, for instance, from being utilitarians in the lever version of the trolley dilemma to being deontologists in the bridge version of the same problem—because we are literally of two minds when it comes to ethical decision-making.

The basic idea is that our cognitive processes (broadly speaking, our ability to think rationally) are engaged in utilitarian ethical judgment, while our emotional responses (our "gut feelings," our intuitions) enable deontological judgment. This concept creates an interesting situation, considering that philosophers think of the two types of ethical theory as logically distinct: thus, we may end up with irreconcilable and contradictory judgments depending on whether one form of judgment or the other takes over in our brains.

What is the evidence for Greene's dual-process theory? Perhaps the earliest clue came with the famous case of Phineas Gage, a nineteenth-century railroad construction foreman who survived the freak accident of a long metal rod passing through his head. Much of Gage's left frontal lobe was destroyed, but this damage did not result in any obvious impairment in his cognitive reasoning compared to before the accident. What did change, however, was his social behavior: suddenly he found it difficult to control his impulsive and emotional reactions. This was the earliest suggestion that the areas of the brain affecting cognition are at least partially different from those controlling emotions, and that it is possible to disrupt (in this case, by accident) the balance between the two.

In the 1990s, research conducted by neurobiologist Antonio Damasio's group zeroed in on a more specific area of the brain, the ventromedial prefrontal cortex (VMPFC), to show that patients with damage there made bad decisions when it came to risk assessment, significantly underestimating the risk associated with certain simulated scenarios. The patients responded normally, however, to tests measuring their ability to engage in moral reasoning; the problem seemed to be caused by the inability of their brains to generate the feelings that normally help guide most of us in analogous situations. Interestingly, studies on the neurobiological underpinning of psychopathy also show a connection with the VMPFC (among other areas of the brain): apparently, psychopathic behavior can be generated by a reduced functioning of the amygdala (the same area that lost its cognitive "brakes" in Jim Fallon's brain), which in turn may be caused by a malfunction in—you guessed it!—the VMPFC. One of the intriguing consequences of psychopathic breakdown of normal brain activity is that psychopaths don't seem to be able to make the distinction that comes easily to most of us between moral rules and arbitrary rules of conduct (such as etiquette-related ones). For them, all rules are arbitrary conventions and can therefore be ignored at will. In a sense, a psychopath is the ultimate moral relativist.

Of course, neurobiological studies focusing on exceptional situations—be they freak accidents or socially deviant individuals—can tell us only so much. Is there evidence for Greene's dual-process theory from more standard situations that affect all of us? Indeed there is. In a study carried out by Greene's group, subjects were presented with what the researchers referred to as a "high-conflict personal dilemma"—something along the lines of the various versions of the trolley dilemma, for instance. The trick was that some of the subjects were simultaneously asked to engage their attention with an unrelated (and morally neutral) cognitive task, such as detecting when the number 5 was presented to them in the midst of a string of numbers. The idea was to cause a simple interference with cognitive moral processing by diverting some cognitive resources to another problem. The dual-process theory would predict that utilitarian moral judgment should be partially impaired by this interference, but not deontological judgment. And that's exactly what the researchers found! It is as if one of the moral channels of the brain shares bandwidth (so to speak) with functions like calculations and identification tasks and the more we are engaged with the latter the worse we do with the former. In addition, we have seen that the experiment can be done in reverse: researchers can interfere with subjects' deontological judgment simply by altering their emotional state in an unrelated fashion—for instance, by exposing them to noxious odors. And of course, all of these findings have more than just scientific import: imagine the endless possibilities for willful manipulation of juries by unscrupulous attorneys bent on shifting the balance of jurors' moral compass toward either a utilitarian or a deontological extreme.

The dual-process theory is also consistent with what we do when faced with very different sorts of moral judgments—those that have to do with the concept of justice (a subject to which we return in Chapters 14 and 15). Consider the following, not too hypothetical, scenario: you have one hundred kilos of food available to be distributed to a population affected by famine. However, it takes some time to deliver the food, and this will cause about twenty kilos to spoil and become unusable. If you choose instead to deliver the food to only half of the population, the spoiled amount will decrease to five kilos. What do you do? If you choose to send more food to only half of the population, you are giving priority to the efficiency of your aid program, but if you still try to deliver to the entire population, despite the greater loss of food, then you are prioritizing fairness over efficiency.

This is precisely the sort of conundrum explored by Ming Hsu and collaborators, who presented subjects with a set of scenarios in which fairness and efficiency could be manipulated independently of each other, and who also obtained brain scans of the participants to figure out not only what they would decide to do under each scenario, but which parts of their brains were involved in the decision-making process. They found that three areas contribute to weighing issues of justice: the putamen is the part that responds to issues of efficiency, the insula is involved with judgments of inequity, and the caudate-septal subgenual region essentially mediates between the two to come up with a unified judgment once the person has considered the relative importance of equity and efficiency in the given situation. Looking at these results, it is hard to resist the conclusion that human beings come equipped with a sophisticated "moral calculator," in much the same way as we are endowed with brain machinery that enables us to learn the complex rules of just about any language during the first few years of our existence.

What is particularly interesting about these results in light of Greene's dual-process theory is that the insula (the inequity-encoding region) is also known to be part of our emotional system; the putamen (the efficiency-encoding region) is involved with the brain's reward system (it is sensitive to dopamine, a natural reward drug produced by our neurons), which in turn has been demonstrated to be linked with feeling good about both charitable giving and punishment of free-riders; and finally, and most revealingly, the area integrating these two functions, the subgenual, has been implicated in trust and social attachment. In other words, it looks like a socially well adjusted person has to constantly weigh issues of fairness and efficiency and that three distinct but interconnected areas of the brain help us do just that. Hsu and his collaborators conclude:

> More broadly, our results support the Kantian and Rawlsian intuition that justice is rooted in a sense of fairness; yet contrary to Kant and Rawls, such a sense is not the product of applying a rational deontological principle but rather results from emotional processing, providing suggestive evidence for moral sentimentalism.

We will get to both Kant and John Rawls (1921–2002) in due time, but if you read this carefully you will find that it is an example of scientists attempting to override philosophy based on experimental results, a violation of Hume's separation between is and ought. As I argue at several junctures in this book, however, this counterposition between science and philosophy is misguided and not particularly fruitful. A more interesting reading of these results is that humans have a built-in emotional sense of fairness analogous to the one advocated by philosophers like Kant and Rawls. But this doesn't mean that, just as with any other biological instinct, rational discourse and learning cannot improve on what mother nature gave us.

Although Greene's dual-process theory is beginning to look like a good way to think about the relative roles of reason and emotion in moral judgment, it is not without its critics. Bryce Huebner of Tufts University, Susan Dwyer of the University of Maryland, and Marc Hauser, formerly of Harvard, have pointed out the obvious problems: from a correlation between emotions and ethical decisions it doesn't follow that the first causes the second; it may just as easily be the case that certain decisions of moral import cause us to experience specific emotional reactions. Huebner and his collaborators do not, of course, deny that emotions are an integral part of the psychology of moral decisions. For instance, it is hard to ignore the fact that the emotions of guilt and shame not only are felt after certain actions but are powerful factors preventing the recurrence of such actions. Still, their paper presents not one but five different models of what they call "the moral mind." Interestingly, four of the five models are associated with the name of a philosopher because each of these four models reflects a well-known type of moral philosophy. Let's take a quick look.

The first possibility considered by Huebner and his colleagues is what they call a "pure Kantian" model: just as philosopher Immanuel Kant thought, in this model Reason influences Emotion, and this in turn generates moral Judgment (thus the causal chain looks like R > E > J). Alternatively, a "pure Humean" model is characterized by the fact that Emotion gets the process started, generating Judgment, followed by our ability to come up with Reasons why we made that judgment (E > J > R). The third possibility, not surprisingly, is a hybrid Kant-Hume model, where both Reason and Emotion interact to yield moral Judgment (E,R > J); this, of course, is essentially a restatement of Greene's dual-process theory. A fourth model is termed by the authors "pure Rawlsian" because it is based on John Rawls's ideas about justice as fairness (to be discussed in Chapters 14 and 15); here moral Judgment comes first (the result of an Analysis of possible Actions), and both Reason and Emotion are deployed to justify it and to act on it (AA > J > E,R). Finally, a "hybrid Rawlsian" model enlists Emotion to carry out action analysis, which then leads to Judgment and finally to the articulation of Reasons in its support. (This would look diagrammatically a lot like the pure Humean model, except for the added interaction between Emotion and Action Analysis: AA/E > J > R).

The interesting point raised by Huebner and his colleagues is that the current empirical evidence does not conclusively discriminate between the five models; indeed, the fifth had not even been articulated in print until they published their paper. Things, therefore, are a bit more complicated than what I have already outlined, though I remain convinced that some version of the Kant-Hume model (that is, a dual-process model) is the one currently favored by the totality of the available evidence.

There is another fairly big set of caveats that an intelligent user of scientific investigations into the functioning of the brain has to keep in mind. As pointed out by Kristin Prehn and Hauke Heekeren, studies like the ones we have examined so far (and to which we return in Chapter 16 when we look at how the human brain treats the concept of gods) are typically based on very small sample sizes, for the good reason that it is still very expensive to put people in an fMRI machine. Although this problem will presumably recede with technological advancements, there are other issues as well. To begin with, when researchers publish those pretty (and pretty convincing!) colored images of brains, pointing to which areas "light up" in response to a particular type of task, we need to realize that these images are statistical composites—they do not show us a particular brain of an individual human being, but rather a sophisticated statistical average across the whole sample of subjects. Moreover, the computer-highlighted spots are not strictly speaking areas where brain activity is higher, but rather locations where the blood flow peaks: the idea is that if the blood flows at a particularly high rate in a certain anatomical area, then oxygen is being exchanged with the underlying tissue and this in turn is the result of increased biological activity by those cells (which need more oxygen to metabolize more efficiently). As a result, then, the images we see in published papers are indirect statistical estimates of brain activity, not actual photographs of it.

There is a more subtle limitation of fMRI studies, known as the *non-interactivity assumption*: it is simply not possible to isolate what is going on in the brain when we do one particular thing only—say, think about the trolley dilemma. That's because the brain does all sorts of other things at the same time, so that we need a way to isolate the "signal" from the focal activity we happen to be interested in. This is done using so-called subtraction logic, another statistical method by which background brain activity is accounted for and eliminated so that the signal pertaining to the task we are interested in emerges more clearly. But the fundamental assumption of subtraction logic is that one can simply add and subtract (as the name implies) different brain activities because they are not interdependent. The problem is that this assumption of noninteractivity between different brain functions is almost surely wrong. We simply don't know how to compensate for the fact that, for instance, the limbic system and the cortex are functionally and anatomically integrated, so that it really isn't possible to separate the "emotional" (limbic) from the "rational" (cortex) activity—they are mixed together in any normally functional human brain.

This is not to say that neurobiology isn't teaching us a lot about how the brain works when it comes to moral decision-making (or anything else, for that matter), but we should remember that, as always in science, what current research tells us should be taken as only provisionally true and that it is likely to be superseded (and occasionally overturned) by better methods and more sophisticated thinking. Still, now that we have some appreciation of what the brain does when thinking about morality, we need to face the even broader question of why human beings have a moral sense to begin with. Why is it that we seem to have a strong instinct to consider some notions "wrong" and others "right"? To get a grip on that question, we need to turn from neuroscience to evolutionary biology.

CHAPTER 4

THE EVOLUTION OF MORALITY

Never let your sense of morals get in the way of doing what's right.

—ISAAC ASIMOV

DURING THE TIME MEN LIVE WITHOUT A COMMON power to keep them all in awe, they are in that condition which is called war; and such a war, as if of every man, against every man.... To this war of every man against every man, this also is consequent; that nothing can be unjust. The notions of right and wrong, justice and injustice have there no place. Where there is no common power, there is no law, where no law, no injustice. Force, and fraud, are in war the cardinal virtues. No arts; no letters; no society; and which is worst of all, continual fear, and danger of violent death: and the life of man, solitary, poor, nasty, brutish and short." These famous words are found in Chapter 12 of *Leviathan* (1651), the political masterpiece by Thomas Hobbes (1588–1679). They reflect a bleak view of human nature, and one that many people think has been vindicated by the modern theory of evolution. Hobbes was not necessarily implying that there had been a moment in human history when people drew up a "social contract" that got them out of their brutish state of nature (and he was certainly not thinking about evolution, given that he was writing more than two centuries before Darwin), but his underlying idea is that morality and justice are latecomers in human history, and that it is only the power of the state that keeps us from sliding back into a war of all against all.

This possibility that morality and civil behavior are a "veneer" precariously imposed on a fundamentally nasty biological nature is entertained by a number of prominent biologists, up to modern times. Thomas Henry Huxley (1825–1895)—known as "Darwin's bulldog" for the persistence with which he defended the newly proposed theory of evolution by natural selection—famously wrote in his *Evolution and Ethics* (1894):

> The practice of that which is ethically best—what we call goodness or virtue—involves a course of conduct which, in all respects, is opposed to that which leads to success in the cosmic struggle for existence. In place of ruthless self-assertion it demands self-restraint; in place of thrusting aside, or treading down, all competitors, it requires that the individual shall not merely respect, but shall help his fellows. . . . Laws and moral precepts are directed to the end of curbing the cosmic process.

Along similar lines, Richard Dawkins, in *The Selfish Gene* (1976), wrote: "Be warned that if you wish, as I do, to build a society in which individuals cooperate generously and unselfishly towards a common good, you can expect little help from biological nature. Let us try to teach generosity and altruism, because we are born selfish." And finally, the influential biologist George Williams (1926–2010) wrote in a technical journal of philosophy in 1988: "I account for morality as an accidental capability produced, in its boundless stupidity, by a biological process that is normally opposed to the expression of such a capability."

If these eminent thinkers are correct in the gist of their argument, then there is no sense in talking about the evolution of morality. Indeed, morality is—in this view—the most antievolutionary of all phenomena: natural selection favors those who look out for number one, the niceties of justice and altruism be damned. But now consider the following scene: Mark accidentally slips and ends up in a stream; not knowing how to swim, he begins to drown. Jean jumps in to help him, even though she herself can't swim and almost gets killed during her heroic attempt; finally, some bystanders intervene and save both Mark and Jean. Or this one: Robert is being tortured by starvation—he can have food if he agrees to send an electrical shock to his comrade Steve, whom he can see across a glass panel. But Robert endures the famine imposed on him to avoid hurting his friend. You might have thought I was referring to common acts of human decency, but it turns out that the first example concerns two chimpanzees (the "stream" was actually a zoo's moat, and the names were made up), while the second one is about two rhesus monkeys (names were also made up, and the situation was experimentally imposed by humans, not an accident, as in the case of the near-drowning).

If the view that morality is a thin behavioral veneer imposed by civilization on otherwise brutish instincts is correct, these and many other examples of what we would consider ethical behavior in nonhumans make precisely no sense at all.

Human beings have a long history of desperately trying to separate themselves from the rest of the animal world; to justify our sense of specialness, we contend that, if not created directly by gods, we at least represent something qualitatively new with respect to the rest of nature. But almost invariably these lines drawn in the evolutionary sand have been obliterated by further discoveries about what happens every day in the lives of our primate cousins. The idea that only human beings are moral animals is just one of the latest casualties of more careful comparative research into primate behavior.

If we truly want to understand our own lives, we need to understand where our powerful sense of right and wrong comes from, and to do that, as primatologist Frans de Waal has pointed out, we need to realize that some of the fundamental building blocks of a moral sense can be found in a number of animal species. Indeed, the closer we come, evolutionarily speaking, to *Homo sapiens* the more obvious the parallels become. Let us start with the very basics: a common misconception, as we have seen, is that evolution is a matter of fighting for your own survival and reproduction because nature is "red in tooth and claw," to use the memorable phrase coined by English poet Alfred Tennyson (1809–1892) in 1849 (who was writing after the publication of one of the pre-Darwinian books on evolution, *Vestiges of the Natural History of Creation*, and who did not mean it as an endorsement of the theory). But beginning with Charles Darwin himself (1809–1882)—who wrote the first book on the evolution of emotions in humans and other animals—biologists have tried to explain how apparently altruistic behaviors could have naturally evolved. Three building blocks of a moral sense have been elucidated by both empirical and theoretical research carried out largely during the second half of the twentieth century, and they are found not just in primates but in a variety of other animal species.

The most basic building block of altruism—and by far the most widespread—is what biologists call "kin selection." William Donald Hamilton (1936–2000) is the researcher whose name is most clearly associated with this concept, as he did the groundbreaking mathematical work that allowed biologists to truly grasp the phenomenon. But we are all very familiar with it: kin selection results in the sort of instinct that makes us give up food, sleep, and wealth, and occasionally even put our own lives at risk, for the sake of a relative, especially a close relative. (Indeed, as geneticist J. B. S. Haldane [1892–1964] famously quipped when asked whether he would give his life to save a brother, "No, but I would to save two brothers or eight cousins," reflecting the biological fact that we share half our genes with our siblings and about one-eighth of them with our cousins.) Natural selection, the argument goes, would favor any behavior that maximizes the chances of passing your genes to the next generation, regardless of whether those genes are actually inside you or in one of your relatives. It is Hamilton's ideas about kin selection that were popularized by science writer Richard Dawkins with his metaphor of "selfish genes."

At this point you may feel a bit outraged by my attempt to reduce human morality to mere biological strategies that allow genes to win their race to the next gene pool. Even more, you may feel that I am not talking about morality at all, because kin selection is present in ants and other social insects, and it certainly doesn't account for the powerful human emotions associated with our love for (and protective behavior toward) our children. But hang on, we just got started.

The second building block we need to begin to understand morality as a natural phenomenon is called "reciprocal altruism." Here the main ideas were developed in different contexts by biologist Robert Trivers and mathematical psychologist Anatol Rapaport. Trivers pointed out that natural selection would favor an altruistic behavior (that is, a behavior that has a cost to the altruist and benefits a nonkin receiver) if there is the expectation of the other actor returning the favor at a later time. There are astounding examples of this in nature, perhaps the most famous one being that of vampire bats. Vampire bats are pretty much what they sound like: their diet is hematophagous—they feed on the blood of other animals. (The common vampire bat makes night raids to feed on the blood of mammals, including humans, while other species feed on birds.) Bat metabolism is so high that if a vampire bat does not feed two nights in a row it dies of starvation. Interestingly, within colonies of these creatures, individuals are regularly seen engaging in what is clearly a high-cost behavior: feeding those whose hunting didn't yield fruits on a particular night. However, a system of reciprocity is established because any particular donor bat may soon find itself in need of help from a bat it has previously helped.

The scenario is actually more complicated than this, since a good number of bats in the same colony are related to each other, which makes this a borderline case between kin selection and reciprocal altruism. But that is the point of naturalistic theories of morality: the phenomenon evolved by intergradations from simple cases of selfishness and kin selection to more complex social behaviors that begin to resemble what humans call altruism. Besides, the case of vampire bats is certainly not the only empirical evidence for reciprocal altruism in nature; another example is the warning call produced by some birds to alert their companions to the approach of a predator (which potentially exposes the caller to direct attention by the predator).

Rapaport's work was of an entirely different nature, and it is only partly related to reciprocal altruism as defined in biology, but it is crucial because it establishes the theoretical soundness of conditionally altruistic behavior among nonrelatives. Rapaport famously won a competition for strategic games organized in 1980 by political scientist Robert Axelrod by entering a program characterized by just four lines of code that embodied the general strategy of tit-for-tat: behave cooperatively to whoever you encounter next, unless he doesn't, in which case you should beat the crap out of him and move on. (I am paraphrasing here.) Turns out that tit-for-tat is both a simple and highly efficacious approach to cooperation within a social setting, and in some versions it is flexible enough to allow the agent to resume cooperation with previously uncooperative subjects, if they change their stance toward the agent. The analogy to a computer-theoretical description of office politics, one can't help but think, is not that far-fetched.

But things become really interesting when we consider one further step in the evolution of morality: "indirect reciprocity." Here the idea is that in certain groups simple reciprocity may not work because the group is large enough, or the number of encounters between the same two agents is small enough, that there is ample room for cheating. For instance, I ask for your help but then successfully shy away from returning the favor because we rarely meet or because the supply of other suckers out there whom I can take advantage of is ample. This situation can be overcome under conditions in which natural selection favors a more sophisticated type of cooperation based on indirect reciprocity. Although the mathematics is more complex, and the number of possible scenarios much larger than in the cases of kin selection and direct reciprocity, the basic idea is simple: social exchanges between two actors are observed by some members of the group, and these observations lead to the formation of each actor's reputation with respect to their social behavior. If someone consistently cheats, his reputation will plummet and he will find it increasingly difficult to receive help, even from people he has not cheated (yet). On the contrary, if someone is reliably helpful, his reputation will get him greater access to help from other members of the group, who are ever more reassured that he will return the favor, should the occasion arise.

Indirect reciprocity is perhaps the building block that really begins to feel like a component of morality as we understand it. Indirectly reciprocal altruism is observed in several other species of primates, which has led biologists to suggest that some of our evolutionary cousins have an innate sense of what humans call "justice." There are several fascinating consequences of the idea of reciprocal altruism, beginning with the necessity for a species that practices it to develop sophisticated cheating-detection and communication mechanisms. You may be familiar with a primary mechanism operating in human populations: gossip. A large percentage of our everyday conversation is one form or another of gossip: it conveys information about other members of our group based on our own (or someone else's) observations of their behavior. This is a crucial component of reputation, and it is the ability to build a reputation within the group that is fundamental for indirect reciprocal altruism to work. Of course, gossiping introduces a whole other level of potential cheating, since it is possible to willfully undermine someone's reputation by lying about what she did—but this sort of behavior, which requires a sophisticated mechanism of communication, such as human language, brings us into familiar ethical territory.

Our understanding of the evolution of morality is not based only on theoretical models and on the strange behavior of distantly related species like vampire bats. Frans de Waal's group at Emory University, as well as other primatologists interested in cooperation and altruism, have repeatedly documented a range of behaviors in our close evolutionary relatives that we unquestionably call "moral" when they are observed in our own species. For instance, food sharing with nonrelatives, even when the animal has the option of monopolizing the food, is found among a large number of primates, including siamangs, capuchin monkeys, orangutans, and the two species most closely related to us, chimpanzees and bonobos. There are also differences among primate species that, again, are consistent with the idea that social and altruistic behavior evolves in response to the particular social ecologies of individual species: for instance, chimpanzees, but not macaques, show a type of retributive behavior in which they intervene on behalf of another member of the group who has been attacked or has developed a conflict with a third party.

Not surprisingly, some of the most humanlike social behavior is found among chimpanzees. Females of that species have been observed mediating reconciliation by extending a hand to a rival, an act often followed up with mouth-to-mouth kissing—perhaps the sort of friendly gesture that humans might benefit by copying from their less socially sophisticated relatives. In the same vein, female chimpanzees mediate the relationship between previously fighting males, often approaching one of them, kissing him (again!), and then gently nudging him toward his former rival. This works more frequently than not, and the two males are later observed grooming each other, seemingly reconciled. Indeed, chimpanzees even indulge in groupwide celebrations after a successful reconciliation attempt.

Most people—including philosophers—think that human morality is built on our ability to empathize, to imagine ourselves in someone else's situation and understand at a deep level what the other person must feel in a certain circumstance. The recent discovery of so-called mirror neurons in both humans and other primates (they are found in birds as well) suggests the possibility that empathy evolved out of the ability to imitate others' behaviors. These neurons fire both when we act in a certain manner and when others act in the same way, clearly making it possible for us to mimic what other members of our species are doing—a fundamental component of our stunning ability to learn. But mirror neurons are probably also involved in what philosophers call our "theory of mind"—our tendency to attribute thoughts and feelings to others by analogy with the thoughts and feelings we ourselves have. From there the step to developing empathy is short.

If this sounds a bit too theoretical and neurobiological, then consider primatologist Jane Goodall's research on chimpanzees: she has documented behavior that is hard not to describe as empathic. When a male is attacked, he will often run screaming in obvious distress to a companion, at which point the two will embrace and scream in concert. Or the distressed male may seek his mother and simply hold her hand. This sort of sophisticated display of emotions is found in the great apes (of which humans are a member), but not in monkeys or other primates, again pointing toward the relatively recent evolution of moral behavior in animals with complex brains and social systems.

But a reasonable objection may be raised: what has any of this to do with true morality, with the conscious sense of right and wrong, with the strong feeling of indignation that humans experience when faced with a situation they consider unjust? Let us start with that "strong feeling" and then work our way to the "conscious sense" part. Much scientific research on moral decision-making—be it of the neurobiological kind (Chapter 3) or the evolutionary variety (this chapter)—points to the conclusion that morality began with automatic social behaviors like kin and reciprocal altruism and eventually generated the set of emotional responses that underlie moral behavior in many primates, including humans. Our ingrained moral sense, in other words, is at bottom a powerful emotion. This should not surprise us: whenever nature wants to convince us to do or not do certain things, it relies on the universally powerful sensations of pleasure and pain. It is a bad idea not to notice that you are bleeding from a deep cut, so natural selection has favored the evolution of nerve cells whose strong pain signals force you to pay attention to the wound, thus possibly saving your life. At the opposite extreme, consider sex. From the point of view of natural selection, it is necessary to convince us to spend time and resources seeking a mate, courting her (or him), and, depending on the species, sticking around to raise the resulting offspring. Hence the evolved pleasures of having sex, with the accompanying neural structures that allow us to enjoy such pleasures (not to mention the natural hormonal high we get from a sense of attachment to our children).

Seen this way, therefore, not only is there no contradiction between a naturalistic explanation of the origin of morality and our strong sense of certain things being right or wrong, but it becomes apparent that it is the evolutionary necessity of moral judgments (in complex social animals) that explains our psychological reactions to perceived injustice. This is true even at the level of the most foundational building block we have examined in this chapter, kin altruism. It is the result of the evolutionary necessity to protect and perpetuate our own genes, regardless of whether they are inside us or our close relatives. In order to achieve this, nature equipped us with strong feelings of protection toward our loved ones, particularly our own offspring (and, in diminishing gradations, toward more and more distant relatives). In other words, there is no contradiction between the biological mechanisms that led to the evolution of moral judgment and the powerful psychological feelings we experience when we engage in such judgments. Indeed, the latter wouldn't exist without the former.

But is that all there is to morality? Is it just a powerful emotion that resulted from natural selection aimed at protecting our own genetic interests? That may be the view of cognitive scientist Jonathan Haidt (we met him in Chapter 2 while talking about trolley dilemmas), who has pointed out that the emotional underpinning of morality must have been in place before our split from the chimpanzee lineage, some 5 to 7 million years ago, and indeed probably earlier than that. Thus, again according to Haidt, the concept of morality as something we think and talk about and try to understand and improve is a very recent newcomer in the evolution of our species—we could not have had such a concept before the evolution of language, a time frame that moves us to just the last 100,000 years. Moreover, if one wishes to talk about morality in philosophical (rather than religious or intuitive) terms, then we are restricted to just the last two or three millennia.

Haidt's conclusion is that when we engage in moral reasoning we are just "confabulating"—we are rationalizing the courses of action we prefer because of our animal instincts. This conclusion, however, is a bit too grim and, especially, too quickly reached. It is somewhat analogous to saying that since language evolved by natural selection, probably to allow social interactions and to coordinate hunting efforts, then the literary achievements of Shakespeare or Dante, the sophisticated way in which modern humans communicate across the globe, the products of literature, philosophy, and science itself are just confabulations, icing on the evolutionary cake. That conclusion would mistake the explanation of the *origin* of something (be it morality or language) with the further *development* of that same thing as a result of human effort and ingenuity.

Instead, we can see the human ability to reason through moral quandaries as a much-needed improvement on our natural moral instinct—just as modern languages, with their complex grammars and vocabularies, are a huge improvement over whatever our Pleistocene ancestors had. Philosopher Peter Singer makes a similar argument with his concept of the expanding circle:

> If I have seen that from an ethical point of view I am just one person among the many in my society, and my interests are no more important, from the point of view of the whole, than the similar interests of others within my society, I am ready to see that, from a still larger point of view, my society is just one among other societies, and the interests of members of my society are no more important, from that larger perspective, than the similar interests of members of other societies. . . . Taking the impartial element in ethical reasoning to its logical conclusion means, first, accepting that we ought to have equal concern for all human beings.

Singer's point is that we can reflect on our biologically based moral instinct and begin to resolve the contradictions that it may entail through reason, which results in an ever-expanding circle of moral concern that eventually includes not only all of humanity but other species as well. (Singer was one of the founding fathers of what became the animal rights movement, even though as a utilitarian he doesn't really believe in rights per se.) This may sound somewhat strange because we are not used to thinking of morality as a mix of biology and philosophy, but the same idea applies to other human endeavors, beginning of course with science. Science is sometimes described as common sense writ large. That's an oversimplification, but the kernel of truth is that science is based on humans' natural ability to gather empirical data about the world through their senses and to draw conclusions about how the world works through their brains' capacity to process that information. Science greatly expands the scope of this natural empiricism by providing us with sophisticated technological tools to augment our senses (microscopes, telescopes, particle accelerators), as well as trained reason to augment our thinking (epistemology, methodology, mathematics). In this sense, moral reasoning is to moral instinct what scientific investigation is to raw observation and intuition; in other words, we come to a better understanding of morality by studying it scientifically at the same time as we improve our moral judgment through philosophical reflection.

CHAPTER 5

A HANDY-DANDY MENU FOR BUILDING YOUR OWN MORAL THEORY

Those are my principles, and if you don't like them . . . well, I have others.

—GROUCHO MARX

MEANING IN LIFE IS A COMPLEX MATTER THAT DEPENDS on circumstances, friends, family, career, and accidents of birth (as in both our genetic makeup and when and where we happen to be born). But, unquestionably, morality plays a crucial role in how we see ourselves and others, which is why we have spent some time here first investigating how the human brain makes moral decisions (Chapters 2 and 3), and then gaining some hint as to why we are moral beings to begin with (Chapter 4). That's all well and good, you might say, but what am I supposed to do about it? Assuming that moral instincts come from evolution, that our brains conduct moral reasoning by a complex combination of logical reasoning and emotional input, and even that gods have nothing to contribute to the issue regardless of whether they exist or not (we'll get to that in Chapter 18), how do we then come up with a reasonable understanding of morality upon which to build the foundations of a meaningful life? Enter the handy-dandy morality menu!

We'll proceed in two steps. First, we will answer what is sometime referred to as the "metaethical" question (the question we temporarily set aside in Chapter 2): if there is no absolute source of morality (like a god), how do we avoid sliding into "anything goes" moral relativism? (The good news is, we don't!) In the second step, we will take a look at the three major ethical systems competing for your vote: deontology (rule-based ethics), consequentialism, and virtue ethics. Finally, I will suggest that, up to a point, we can in fact mix and match the insights from these three major views of ethics to build a custom-fitting, yet not arbitrary, ethical view for ourselves.

Metaethics is that branch of philosophy that asks the fundamental question: how do we justify ethical reasoning at all? We shall see toward the end of the book that the most common answer—that morality is a god-given gift to humanity—simply won't do, for very solid philosophical reasons. A good number of people take this to be an admission that ethics is therefore a matter of taste: I prefer dark chocolate and you go for the milk variety (really?); I prefer that young girls' genitals not be mutilated, and you don't mind it so much. Who's to say what's right and what's wrong? I cannot make an argument that preferring milk over dark chocolate is irrational (though it does seem very strange to me), but I can make a very substantial argument that—political correctness about other cultural norms notwithstanding—some actions are just wrong, across the board of human experience.

Discussions of metaethics can get really complex, since philosophers tackle the problem by taking a variety of approaches. Still, what is needed for our purposes here is to understand the peculiar role played by "facts"—our empirical observations about human behavior (and hence science)—when it comes to issues of "values"—our ethical choices. In the discussion on the relationship between science and philosophy at the beginning of the book, we encountered David Hume's naturalistic fallacy: the idea that one cannot simply move from a matter of fact (what is) to a matter of value (what ought to be). There are plenty of natural things that are simply not good for us—poisonous mushrooms come to mind, for instance. Similarly in ethics, just because it is natural (that is, instinctive), say, for people to distrust outsiders, it doesn't follow that we should treat immigrants any differently from the way we treat native-born citizens.

Recently a spate of claims have been made by scientists, and particularly neurobiologists, that research into the structure of the brain can solve all sorts of philosophical problems, from the issue of free will (see Chapter 9) to the question with which we are presently occupied: morality. Perhaps the most comprehensive scientific attack on moral philosophy is the one mounted by author Sam Harris in his book *The Moral Landscape: How Science Can Determine Human Values*. Harris dismisses the entire philosophical literature on ethics in a footnote at the beginning of his book, on the grounds that "every appearance of terms like 'metaethics,' 'deontology,' . . . directly increases the amount of boredom in the universe." That assertion, needless to say, is simply not serious scholarship (not to mention that one could level the exact same "criticism" at every appearance of terms like "fMRI," "parietal lobe," "axon," and so forth—you get the drift). Moreover, again in a footnote, Harris helps himself to such a broad definition of science that one can make a reasonable argument that he is actually talking about scientia (which, as we saw at the beginning of the book, includes both science and philosophy and is what I call sci-phi), since he does "not intend to make a hard distinction between 'science' and other intellectual contexts in which we discuss 'facts.'" But then he isn't talking about the "science" that, in the actual sense practiced by scientists and understood by most of the public, can help us determine our values. Harris is playing a game of bait-and-switch with his readers.

The more substantial criticism of Harris's endeavor (and similar others), however, is that he simply does not seem to acknowledge or understand the distinction between facts and values. This is most comically on full display in a passage where he reports on his own neurobiological research on the fact/value distinction:

> First, belief appears to be largely mediated by the MPFC [medial prefrontal cortex], which seems to already constitute an anatomical bridge between reasoning and value. Second, the MPFC appears to be similarly engaged, irrespective of a belief's content. This finding of content-independence challenges the fact/value distinction very directly: for if, from the point of view of the brain, believing "the sun is a star" is importantly similar to believing "cruelty is wrong," how can we say that scientific and ethical judgments have nothing in common?

This is nonsense on stilts. To begin with, nobody has ever claimed that scientific and ethical judgments have nothing in common, from the point of view of the brain. More importantly, it most certainly does not follow from this that facts and values are the same sort of thing, only that the brain deals with both in the same areas. (By similar reasoning, since the same areas of the brain respond to having sex and thinking about having sex, it would follow that the two experiences are one and the same. Another stunning discovery of modern science.)

Finally, Harris's entire project is predicated on the idea that science is the best way to judge the consequences of our actions and to channel them in the direction of increasing human well-being, by which he means an increase in happiness and a decrease in pain. But this project is incredibly philosophically loaded, as Harris is taking for granted a particular consequentialist approach to ethics (utilitarianism), which is far from the only contender in the field of moral philosophy (as we'll see in a minute). Needless to say, he does this with neither a scientific (because it is not possible) nor a philosophical (because it is boring) defense of his assumptions.

My critique of Harris and other neuro-enthusiasts aside, empirical facts are most certainly not irrelevant to ethical judgment. Morality is a human attribute, and it makes no sense to think about it without regard to what sort of animals we are. In an important sense, ethics is about human well-being (I will leave aside the issue of animal rights, for the simple reason that animals would not have "rights" unless there were human beings capable of thinking about such things as rights), so we want to know empirical facts about what increases or decreases our well-being whenever we engage in ethical reasoning. (Incidentally, the very concept of well-being is in fact open to much discussion, both philosophically and in terms of social science research. We'll take a close look at the idea in the last chapter of the book.) Let me put this point another way: morality makes sense only within the context of a group of social animals capable of reflecting on what they are doing and why. A lion that kills another male's cubs when he takes over a harem is not acting immorally, he is just being a lion. Similarly, even a human being could not possibly commit immoral acts if he or she were, say, permanently stranded on a deserted island, because against whom would such acts be committed?

It should be clear from our discussion of the evolution of morality in the last chapter that evolution is where humanity got the foundational building blocks for a moral sense. We can philosophize all we like about what is right and what is wrong, but unless we have a strong natural instinct that makes us care about perceived injustice, all such discussions are literally academic and do not affect our lives. Evolution has equipped us with only a very basic moral instinct that is tailored to work in the situations that historically affected our survival—that is, within small bands of individuals who had to stick together and often defend themselves from aggression not just by other species but by members of *Homo sapiens* who belonged to different tribes. The idea of morality as a system of thought rather than an animal's unconscious assessment of certain social situations relies on the unique ability of humans (so far as we know) to reflect on what we do and why we do it. That, of course, is where philosophy comes into play to show us how to build on our instincts and broaden our conception of what it means for something to be right or wrong.

Consider again the example I mentioned earlier: humans' strong innate feeling that members of their in-group ought to be treated with respect because survival depends on them. We eventually realized that there is no reason why this conditional reciprocity (the "tit-for-tat" strategy we encountered in the last chapter) should not be extended to every other human on the planet, simply on the grounds that human beings are fundamentally the same everywhere, not just in our own neighborhood. What is interesting in making this move is that the underlying reason for our attitude changes in an important manner: natural selection presumably endowed us with a tendency to engage in tit-for-tat with our neighbors and fellow in-group members because such behavior is (demonstrably, by mathematical modeling, as it turns out) the best evolutionary strategy to maximize our own well-being. Recall that tit-for-tat boils down to being nice to your fellow humans unless they are being nasty to you, in which case you retaliate. This is the same behavior that can be observed to this day in our closest evolutionary cousins, the bonobo chimpanzees. But when we enlarge the conditional reciprocity circle to the entire human race, we do so because by reflective reasoning we see that it is the right thing to do. It is not that what other humans do thousands of miles away actually affects us (at least not most of the time); it is that we realize that we would not want, say, torture, mutilation, killings, famine, and so on, to happen to us and there is no rational defense of the idea that we are somehow more deserving than anyone else on the planet.

What we have, then, to simplify a bit, is a two-step process to answer the metaethical question. In the first place, we would not even be talking about morality if it were not for the contingent fact that we are a species of large-brained social animals. Here science tells us the most plausible story of how the rudiments of moral thinking (or more precisely, feeling) came about. Second, our ability to communicate with each other and to critically reflect on our own actions generated the concept that we ought to extend the reach of our moral system to (at least) all other members of our species (and arguably beyond). This is philosophy's contribution to the problem.

And speaking of philosophy, it is finally time to turn to a brief examination of the three major schools of thought vying for the title of moral system of choice, beginning with deontology. The basic idea, as we saw in Chapter 2, is familiar to anyone who subscribes to a religious system of morals: there are rules to follow unquestioningly, and they are to be followed because they spell out what is the right thing to do. The Ten Commandments of the Judeo-Christian tradition are a classic example of a deontological (duty-based) moral system. But wait, did I not just say that gods have nothing to do with morality? Yes, but modern deontological theories are based on philosophical analysis, not on theology. By far the most influential of such systems is the one developed by Immanuel Kant. A deeply religious person, Kant was raised in a Pietist household where his parents taught him to take the Bible literally. But Kant the philosopher, eventually realizing that ethics needs a rational foundation independent of any particular religious tradition, set out to provide that foundation. (Curiously, Kant is also the philosopher who arguably provided us with the best arguments against the existence of God, presented in his *Critique of Practical Reason*.)

In one of his most influential books, *Groundwork of the Metaphysics of Morals*, Kant formulated several versions of his now-famous "categorical imperative," the foundation for his deontological system. One version reads, "Act only according to that maxim by which you can at the same time will that it would become a universal law," and another goes: "Act in such a way that you always treat humanity, whether in your own person or in the person of any other, never simply as a means, but always at the same time as an end." The first version of the imperative should sound very familiar: it is "the Golden Rule" espoused by many religions worldwide (not only Judeo-Christianity but also Buddhism, Confucianism, Hinduism, and Taoism, to mention just a few). The second version of the imperative is a bit more philosophically subtle and wider-ranging than the Golden Rule: Kant is saying that to act morally is to think of others as beings endowed with exactly the same intrinsic value we hold ourselves to have, and never as tools to achieve another goal.

Although this imperative sounds reasonable and certainly noble, it is actually very difficult to practice. For instance, suppose that I do something nice for my friend, or imagine that I give money to a charity to help the victims of a calamity. Naturally, I will feel good about doing such things, and indeed, I may even pat myself on the back for being such a nice guy. According to Kant, however, if I do so I would not be acting morally. My behavior would not be immoral, but since I derived pleasure from it, a pure Kantian could argue that I acted kindly in order to feel better, hence, that I used other people as a means to an end of my own.

Perhaps you can begin to see why Kant developed a reputation for being a strict moralist with little understanding of, or sympathy for, human nature. But of course in order to accept a deontological system of morality we do not have to be as joyless as the famous sage of Königsberg (Kant's natal city, where he spent his life). We can agree that other people have the same intrinsic value as we do, and yet we can still feel okay about our natural tendency toward self-congratulation for a deed well done. Moreover, we can agree that the categorical imperative is a generally good rule regardless of Kant's own stricter interpretation of morality.

Deontological systems do run into some interesting problems, however, outside of Kant's idiosyncratic brand of moralism. Indeed, we encountered one major issue for deontology when we examined the trolley dilemma. We saw that most people agree that it is justified to kill one person to save five by switching a lever that causes the runaway trolley to change course. But in so doing, obviously we do treat the unfortunate victim as a means to an end (saving the other five). Regardless of how ethically praiseworthy that end is, we are in direct violation of the categorical imperative. This sort of problem arises for deontologists because deontology tends to be concerned with intentions (as opposed to consequences) and with universal rules, while sometimes good moral intentions are best served by not following universal rules (so-called situational ethics). For instance, Kant had a serious problem with the idea of lying and argued (correctly) that, if universalized, lying would be an unqualified disaster for society. Yet it is easy to come up with instances where lying is not only acceptable but the right thing to do—as when, in the classic example, a Nazi officer knocks at your door, asking whether you are hiding a Jewish fugitive, and you lie to him, saying that you are not. You can see how the idea of universal imperatives, as attractive as it is initially, quickly leads to significant moral dilemmas.

One way to resolve these problems, as we also saw in discussing the trolley dilemmas, is to turn to a second major school of thought in ethics: utilitarianism. Utilitarianism was the original idea of two influential British philosophers, Jeremy Bentham and John Stuart Mill. Their cardinal concept is the so-called principle of utility: an ethical action is one that furthers the greatest happiness for the greatest number. Subsequent philosophers have elaborated many variations of utilitarianism, producing, for instance, a "negative" version that essentially states that morality is about reducing suffering (as opposed to increasing happiness). This is the principle that

Princeton's Peter Singer extends to other animals as well as human beings, thus providing a serious philosophical underpinning for the animal welfare movement.

This approach to ethics is often referred to in the modern literature as "consequentialism" because it can be thought of as a way to evaluate moral choices in terms of the consequences of our actions (as opposed to the intentions underlying them); this is a radically different approach from deontology, as we have just seen. Consequentialism seems to be better able than deontology to handle situations like the classic trolley dilemma: we ought to switch the lever because the consequences of our action (saving five lives while sacrificing one) increase overall happiness (or at least decrease overall pain) over what would happen if we did nothing (one person survives but five die).

Still, consequentialism itself can be criticized on various grounds. There are two common objections to the idea, both of which present serious problems for utilitarians. The first is that it is not at all clear how far we need to foresee the consequences of our actions—or indeed, whether we are even capable of doing so. Suppose, for instance, that I live in a country where a horrible dictator is oppressing his people, and I decide that the only way out is a revolution. I manage to organize a clandestine operation that, truly motivated by the best ethical intentions, attempts a coup d'état. Alas, the revolution fails, and as a result, not only do thousands of my comrades die, but the dictator clamps down even more on my fellow citizens' liberties, thus enormously decreasing everyone's happiness and increasing their pain. Even though I did my best to do the right thing, the whole enterprise ended up being a disaster. A strict consequentialist could then argue that my decision to start the revolution was immoral. (You might have noticed by now that philosophers engage with some gusto in producing disturbing thought experiments. I wonder what a psychologist would make of that.)

The logical answer to this objection is that we are morally responsible only for the consequences we can reasonably foresee. That sounds pretty straightforward, until we realize that what we can foresee in part depends on how much we are capable of giving our actions (and their consequences) some reflection, as well as on how much reliable information we have concerning potentially very complex chains of events. For instance, let's modify the revolution example: suppose that I was very careless in the whole enterprise and started the revolution out of pure youthful enthusiasm, dramatically overestimating the likelihood of success. Now, of course, I do carry more moral responsibility for the pain and suffering that ensued as a result of my actions. But then again, we often think of young and idealistic people as morally praiseworthy because they at least try to improve things. What actually happens is often beyond anyone's direct control and comes down to what philosopher Thomas Nagel calls "moral luck." This line of reasoning does not necessarily pose a fatal objection to consequentialism, but it highlights how difficult it can be to make a decision based on an evaluation of its possible range of consequences.

A second classical objection to utilitarianism is represented by the so-called transplant problem, a more malicious version of the trolley dilemma, if you will. Imagine that you are a surgeon at a local hospital and five injured people are brought into the emergency room. Each has a lethal lesion in one vital organ—the liver, heart, kidneys, pancreas, or lungs. (The only other vital organ in humans is the brain, and for now at least we don't have the medical know-how to do a brain transplant—not to mention that such a procedure, if possible, would raise a panoply of philosophical questions in its own right!) If you are a consequentialist doctor, it would seem that you should seriously consider forcing a perfectly healthy person who happens to be standing nearby to undergo an operation so that you can transplant his vital organs and save the five lives. This would certainly increase the general degree of happiness, or decrease the general amount of pain, but most of us would unhesitatingly say that anyone who acted that way would be a monster who should be prosecuted to the fullest extent of the law. It seems, then, that there is something flawed about the principle of utility (though, predictably, there are some reasonable counter-objections that utilitarians can mount here).

If neither deontology nor consequentialism quite cut it, is there perhaps a third option? As it happens, there is, and we encountered it already in Chapter 1: so-called virtue ethics. It was first proposed by Aristotle, and in an updated version it's the third major modern contender in the field of ethical theories. The first thing we need to understand about virtue ethics is what Aristotle meant by "virtue." To possess a virtue is to be a certain kind of person, to have a character trait that is morally valuable. For instance, a virtuous person would be honest, and that honesty would reflect not just a natural propensity but also the person's thoughtful recognition that honesty and truth are to be valued; in other words, one is not virtuous in this sense by accident, but because one works at it. The second crucial aspect of virtue ethics is that, unlike deontology and consequentialism, it does not directly address the question of "what is the right thing to do?" but rather deals with the more fundamental issue of "how are we to live?"

As we have seen, for Aristotle a virtuous person can overcome akrasia (weakness of the will) in order to achieve eudaimonia (to flourish). According to virtue ethics, then, human beings need to steer themselves in the direction of virtuous behavior both because that is the right thing to do and because the very point of life is to live it in a eudaimonic way. Interestingly, however, Aristotle also thought that you have to be at least a bit lucky in order to be able to live a eudaimonic life: if you happen to suffer from an extremely crippling disease or live in dire external circumstances, you may not be able to flourish and your very character will be affected by such conditions. While many ethicists (except Nagel) find the idea that luck has anything to do with morality rather bizarre, I think Aristotle got it exactly right in this case, demonstrating his deep understanding of the complexity of the human condition.

Just as in the other two cases, there are of course reasonable criticisms of virtue ethics. The chief one perhaps is that the idea sounds good in general, but is short on specific guidance about how to conduct our lives. There surely are several different ways for a human being to flourish, and Aristotle's suggestion that we steer a middle course among extremes in order to be virtuous may be hard to apply because "the middle" is a large space. Consider, for instance, the example of courage, one of the Aristotelian virtues: too little of it makes a person a coward, while too much of it makes him reckless. But this notion is unlikely to help us when we are trying to figure out whether we should risk our life in a specific circumstance.

Then again, virtue ethicists turn this objection into an advantage: we have seen with both deontology and consequentialism that they run into trouble precisely because they attempt to codify behaviors too rigidly, either according to a particular set of rules (deontology) or by following a simple overarching criterion (consequentialism). Real life is too complex for that, and ethical decisions are indeed difficult to make, which is why virtue ethics' emphasis on character rather than actions or intentions may ultimately be more realistic.

Another objection often raised against virtue ethics is that the whole concept of akrasia is absurd because it does not make sense to say that I may decide to do something against my will (like not going to the gym when I really want to). This objection stems from an overly rationalistic view of humanity, however, and completely misses the psychological dimension of what it means to be human. If you truly do not know what it is like to have to battle the weakness of your will, you are a very peculiar human being indeed (and may be in for a shock when you read Chapters 9 and 10).

As is often the case in both science and philosophy, there is no ultimate answer to the question of how we should conduct our life. But that doesn't mean that there is no answer, or that all answers are about the same. In particular—and despite the possible protestations of professional philosophers— we can put together a reasonable view of the ethical and meaningful life by combining elements of all three major moral theories, and the specific combination does not have to be the same for everybody. For instance, I am particularly sympathetic to virtue ethics because I find the idea of flourishing as a lifelong project attractive, and because I easily recognize my own akratic tendencies. But I am also aware of the power of Kant's categorical imperative, which I interpret just a bit less strictly than that august philosopher. And finally, I also find much value in the consequentialist emphasis on our personal responsibility to make informed decisions because morality has a lot to do with the ramifications of our actions.

A professional moral philosopher will probably object to this project of constructing a morality menu on the ground that some of the ethical concepts we have examined entail mutual contradictions, so that it is not possible to arrive at a coherent system of thought by combining the best of virtue ethics, deontology, and consequentialism. I am tempted to respond as the American poet Walt Whitman (1819–1892) famously did (in *Song of Myself*): "Do I contradict myself? Very well then I contradict myself, I am large, I contain multitudes." There is some wisdom in Whitman's retort. Although we will see in Chapters 14 and 15 how philosophers engage in a very useful type of reflective exercise to increase the internal coherence of their beliefs, there is something to be said for the possibility that human affairs are simply too messy to be interpreted through a rigid, formal, logical approach. Just as in applied science—in medical research, for instance—we sometimes have to make decisions that cannot wait for more data to be accumulated or for the perfect experimental protocol to be devised, so in practical philosophy we may have to live with the best that can be done instead of seeking an unattainable Platonic ideal. Nevertheless, in order to live an ethical and meaningful life we have to think about what we are doing and why, and such thinking is greatly helped by considering what the greatest philosophers of all time have had to say about the human condition. It is then still up to each and every one of us to decide what to make of it all.

PART II

HOW DO WE KNOW WHAT WE THINK WE KNOW?

CHAPTER 6

THE NOT SO RATIONAL ANIMAL

Errors of opinion may be tolerated where reason is left free to combat it.

—Thomas Jefferson

As we have seen when talking about morality, human beings have always desperately tried to differentiate themselves from the animal world. A probably apocryphal but nonetheless illustrative story has it that a Victorian lady, upon hearing for the first time of Darwin's theory that we are related to chimps, commented that even if true one would hope that the news didn't get around because it would be embarrassing. But the news did get around, and science has made it increasingly difficult to find clear-cut differences between us and other animals. We are not the only animals to use tools, for instance, and not the only ones to engage in cultural practices. Still, two things stand out as uniquely human as far as we can see: language (not just communication) and (deliberative) reason. Aristotle, in Book Z of his *Metaphysics*, defines man (today we would say humans) as the rational animal, a definition that acknowledges both a continuity with the rest of the biological world (we are animals, after all) and a sharp qualitative difference that separates us from it. As we shall see in this chapter, however, it may be more accurate to think of human beings as the rationalizing animals, with language—ironically—providing a key tool that confuses our own and other people's thoughts. If we wish to pursue the fulfilled life, then, we need to come to terms with how easily we fool ourselves into thinking what we ought to know better than to think.

It is remarkably easy for our brain to be manipulated into believing that we are making a rational decision when in fact we are doing anything but. I experienced this firsthand a few years ago when I participated in a live demonstration at the taping of the National Public Radio show *Radiolab*. We were first asked to think about the last two digits of our social security number, but not to tell anyone. Then we were presented with an item that we might have purchased in a toy or electronics store and asked how much we thought was reasonable to pay for it. I did not see the connection between the two exercises until the hosts of the show lined everyone up in order of what they were willing to pay for the item and thus showed us that there was a perfect correlation with our social security numbers: the higher the last digits of our social security number, the more we thought it was "reasonable" to pay for the proffered item. This is an example of what psychologists call "priming": once you start thinking about something, even though it is logically unrelated to the task at hand, you take a certain attitude toward the task that is best explained by the priming effect, not by any objective characteristic of the task. This is also why, for instance, interviewers (or people on a date) can be induced to rate a job candidate as "cold" or "warm" depending on whether they were holding an iced or hot drink in their hands while conducting the interview! We should beware not only of job interviewers but also of advertisers, courtroom lawyers, and a host of other people and situations in which our brains can be manipulated without our even noticing.

A related phenomenon is known as "framing," and it has been demonstrated in a variety of circumstances and experimental settings. For instance, research carried out by Benedetto de Martino, Dharshan Kumaran, Ben Seymour, and Raymond J. Dolan and published in the prestigious *Science* magazine shows how easy it is to nudge people toward either a risk-averse or a risk-prone financial behavior, simply depending on how one poses *the exact same problem* to them. De Martino and his colleagues asked their subjects to think about what they would do with a certain amount of money given to them, of which they would be able to keep only a certain amount. People behaved very differently if the problem was framed in terms of "keeping" the money versus "losing" the money: apparently, to our brain it makes a difference whether we are told that we get to *keep* $60 of a $100 gift (thereby losing $40), or that we will *lose* $40 (thereby keeping $60), even though obviously the two situations are logically identical. We are not talking high math here, or complex probability theory, and yet intelligent persons make radically different decisions about equivalent problems simply depending on how the problem is presented to them. This is why the next time you hear about the results of an opinion poll the first thing you may want to ask yourself is: how were the questions framed? The results of the poll might have been very different had the researchers framed the questions in another fashion. (We return to framing in the context of politics in Chapter 13.)

But if we really wish to get a good feeling for just how weird our brain can be in terms of beliefs and how to rationalize them, we need to look into the vast literature on brain injuries, particularly the many studies of delusions. *Delusion* is not just a common derogatory word, but has a technical meaning as well. The fourth edition of the *Diagnostic and Statistical Manual of Mental Disorders*, published in 2000, defines it this way:

> A false belief based on incorrect inference about external reality that is firmly sustained despite what almost everyone else believes and despite what constitutes incontrovertible and obvious proof or evidence to the contrary. The belief is not one ordinarily accepted by other members of the person's culture or subculture (e.g., it is not an article of religious faith). When a false belief involves a value judgment, it is regarded as a delusion only when the judgment is so extreme as to defy credibility.

Besides noting the odd exception that the authors of the DSM-IV made for religious belief (does something cease being a delusion just because a large number of people share in it?), one can also argue that this definition is far too broad. For instance, a good number of Americans do not believe that the earth is billions of years old, "despite what constitutes incontrovertible and obvious proof or evidence to the contrary." This isn't just a result of religious fundamentalism, as it is often simplistically assumed, but stems also from the fact that the "incontrovertible and obvious" proof appears as such only to people technically trained in biology, geology, or physics. Still, there are some peculiarly instructive cases that would qualify as true delusions by pretty much anyone's standards—except, of course, those of the patients affected.

Take, for instance, the following heart-wrenching description of a patient affected by Cotard's syndrome, which manifests itself in the delusion that one is dead:

> She repeatedly stated that she was dead and was adamant that she had died two weeks prior to the assessment (i.e., around the time of her admission). She was extremely distressed and tearful as she related these beliefs, and was very anxious to learn whether or not the hospital she was in, was "heaven." When asked how she thought she had died, she replied "I don't know how. Now I know that I had a flu and came here on 19th November. Maybe I died of the flu." Interestingly, she also reported that she felt "a bit strange towards my boyfriend. I cannot kiss him, it feels strange—although I know that he loves me."

Cotard's syndrome can manifest itself in a variety of stunning ways: sufferers may be convinced, for instance, that their flesh is rotting, or that they have lost their internal organs. Occasionally, the disease causes the delusion of being immortal. Cotard's syndrome is caused by a disconnect between the area of the brain that recognizes faces and the amygdala, the site of emotions. In other words, sufferers see their own face in the mirror, but do not respond emotionally as if it is their face. The rationalizing brain then kicks into high gear and makes up an explanation to account for the disconcerting sense data: if the person in the mirror looks like me and yet doesn't feel like me, it must be because I'm dead.

Something very similar goes on in a related delusion known as Capgras syndrome. Those suffering from this delusion think that a close person, like a spouse, a friend, or a parent, has been replaced by a look-alike impostor. Again, this outrageous conclusion is reached by a brain that has been confronted with an otherwise inexplicable set of facts about the world as it perceives them—in this case, the disconnect between other people's appearance and what one feels for them. Those suffering from this delusion simply must make up a story to account for what has happened, because that narrative provides them with the illusion that they are in control, regardless of how flimsy the alleged "explanation" actually is.

Perhaps the best evidence that the brain often works more as a rationalizing than as a rational agent comes from classic experiments with "split-brain" patients. All human beings have two hemispheres in their brain, which are anatomically and functionally distinct. This is not unusual in vertebrates, and animals from fish to mammals use the left hemisphere to control everyday behavior while the right hemisphere is more apt to deal with unusual circumstances. (Did you need any further proof of evolution?) In our species, a special structure called the corpus callosum connects the two hemispheres in normal individuals, ensuring continuous communication and coordination between the two halves of our minding organ. In some people, however, the corpus callosum is severed, either because of an accident or, as in most cases, because of emergency surgery to alleviate the symptoms of extreme epileptic seizures. The behavior of these split-brain patients can be followed, both to help them cope with their unusual situation and to help researchers learn more about how each hemisphere works in isolation. The results provide a stunning insight into what must be the normal—and usually unseen—operation of the human brain.

One of the classic experiments of this type was conducted by Michael Gazzaniga's group at Dartmouth College. They took advantage of the fact that one can show images to only one hemisphere, since the right hemisphere controls the left half of the visual field, while the left hemisphere has access to the right half. Moreover, the left hemisphere can communicate verbally, while the right cannot; the right, however, controls the left arm (just like the left hemisphere controls the right arm), so it can respond to questions nonetheless. The experiment began with the right hemisphere being shown the image of a house during a snowstorm while the left hemisphere was presented with a bird's foot. Researchers could communicate with the two hemispheres separately, since the left one controls language, while the right one responds to visual cues. Each hemisphere was then asked to pick the most appropriate image among several to be logically associated with the one it had just been shown. Both hemispheres responded correctly, the right one picking a shovel (to go with the snowstorm), the left one choosing a chicken (to go with the bird's foot). Things got interesting once the experimenters asked the left hemisphere—verbally—why the patient (whose actions had been prompted by both hemispheres, working independently) had picked a shovel and a chicken. Remember that the left hemisphere had no access to the decision-making process of its right counterpart, because of the severed corpus callosum. That didn't stop it from providing an apparently rational, and yet entirely made up, explanation: the shovel had been picked in order to clean the chicken shed!

The irony here is that modern neurobiological research seems to indicate that the right hemisphere is more veridical, sticking to a direct interpretation of the information that reaches it, while the left hemisphere—the one in charge of language—is prone to weaving complex narratives somewhat detached from reality in order to make sense of contradictory information and relieve what psychologists call "cognitive dissonance." Here is a simple example of how this tendency to look for explanations that are more complex than necessary gets us into trouble: it turns out that even rats can beat us at a simple cognitive task! The task consists of figuring out that the lights appearing on a screen are both random (there is no underlying organizing rule or principle) and statistically more likely to appear on the top portion of the screen. Rats and other animals figure out (obviously, not consciously) that the lights have a tendency to appear at the top of the screen and quickly develop an optimized strategy of pushing the appropriate button to obtain a reward. Human beings significantly underperform the rats because they insist on concocting complex theories about the real rule generating the pattern; since there is no such rule, their chance of reward is much lower. It is hard to imagine a more elegant demonstration that overthinking things is not a good idea.

Sometimes our overthinking results in what cognitive scientists call "confabulation." Spontaneous (that is, nonpathological) confabulation occurs when we are pushed to retrieve details of an event that we do not remember. Always attentive to the stress that results from cognitive dissonance, the brain immediately "retrieves" memories that are not actually there, literally making up stories as we go to reduce the distance between what we are told we should know and what we remember. This is how at least some of the so-called repressed memories come out in psychotherapy, occasionally resulting in unjust prosecution of parents accused of child sexual abuse by their own children, even though the "memory" was simply a result of confabulation that the patient's brain concocted to reduce the cognitive dissonance induced by the therapist's probing.

As usual, we learn a great deal about how the brain works when it fails to do so: confabulation can turn pathological, often as the result of damage in the orbitofrontal-anterior limbic system, although it can also be induced by environmental stresses, such as alcoholism. In these cases, not only are individuals absolutely convinced of the stories they produce, but they strenuously defend their version of reality even in the face of obvious evidence to the contrary. For instance, a patient admitted to a hospital in Berne (Switzerland) insisted that he was in Bordeaux (France). When confronted with the landscape outside the window, he admitted that it didn't look like Bordeaux, but immediately added, "I am not crazy, I know that I am in Bordeaux!"

The phenomenon of confabulation is neurologically complex, and the spontaneous form that may affect healthy subjects is significantly different from the pathological form. Nonetheless, what seems to be happening is that something goes wrong with the brain's ability to keep past memories separate from the monitoring of ongoing reality, with the result that information about what is going on gets mixed up in the brain and its "rationalizer" kicks in at full speed to make some sense—any sense—of the jumbled information.

All this evidence from neuroscience may not have shaken your faith in human beings' reasoning abilities that much. After all, we have mostly been talking about pathologies—even though scientists like V. S. Ramachandran maintain that Cotard's and Capgras syndromes, as well as the behavior of split-brain and confabulating patients, are just an exaggerated and easier to study version of what normally goes on inside our minds. The fact, however, is that Aristotle's view of humans as the rational animal is also shaken by more subtle evidence from cognitive science that pertains to how people make everyday judgments about personal and social issues—as, for instance, when we are caught between the necessity for rational decision-making and our tendency to "go hedonic."

A study conducted by Michel Cabanac and Marie-Claude Bonniot-Cabanac presented subjects with a series of questions about social issues ranging from abortion to homosexuality and from climate change to the situation in the Middle East. They were then asked to provide their judgment of a number of possible solutions to each issue. The researchers asked the subjects to rate the possible solutions in two different ways (at different times): first, they were asked to indicate which solutions for each issue felt pleasurable, neutral, or displeasurable; and second, they were asked which solution they would pick if they had the power to enact it.

The results were insightful. To begin with, when subjects were under time pressure they tended to go for the hedonic solution—that is, the answer that made them feel better. However, if given more time and explicitly told to weigh the options neutrally and rationally, subjects tended to change their preferences and were much less likely to pick whatever made them feel good. In their discussion of these results, Cabanac and Bonniot-Cabanac also cite previous studies establishing that when people are in a good mood they are more likely to assess situations rationally, as opposed to hedonically. As it turns out, the human brain also feels pleasure when it engages in logical and rational thinking—for instance, when we are able to figure out the answer to a puzzle—but it predictably feels significantly more pleasure (as measured by the release of brain endorphins) when the choice is hedonic. One can see the results of this type of research as either encouraging or dispiriting. The pessimist could reasonably point to the conclusion that people prefer judgments that make them feel better over more considered ones and that as simple a thing as mood affects our judgments on moral and social issues. The optimist, however, could just as reasonably respond that when people are made aware of these tendencies and given a setting suitable for reflection, they are capable of more sophisticated judgments.

All of this is important to our quest for a meaningful life because such a life depends not only on our ability to exercise the best judgment possible on a variety of issues but also on the overall functionality of our society as a thriving democracy. Which brings me to the troubling research conducted by political scientists Christopher Achen and Larry Bartels at Princeton University. Achen and Bartels were interested in empirically investigating how voters make up or change their mind on important issues and to what extent they use party affiliation as a "proxy" (a kind of shortcut) for their decisions.

The first case study concerned people's perception of how the federal budget fared during President Bill Clinton's first term. Let's begin with the actual data: the deficit was significantly reduced by the end of the term, by about 90 percent. What did people think had happened? Overall, only about one-third of the American public appreciated the fact that the deficit had been reduced, while 40 percent (and 50 percent of Republicans) thought the deficit had actually increased. This finding is particularly discouraging because the amount of the federal deficit isn't a matter of political opinion (unlike, say, the best way to deal with the deficit), and the pertinent information is both easy to comprehend and widely reported by the media. Indeed, Achen and Bartels found the situation to be slightly better when they focused on the 20 percent of voters who were most informed: among these voters, they found, reality had a "pulling effect." Still, they concluded that "reality seems to have had virtually no effect on the responses of people in the bottom two-thirds of the information scale."

The picture got even more complicated when the researchers turned to a second case study that used data from the same period: people were asked whether the economy had improved over the previous year (1996). The fact was that the economy had indeed improved, and a full three-quarters of people knew that. (A comforting percentage, of course, but we are still left with one out of four Americans being willfully ignorant of a basic and very practical fact affecting their lives.) Partisanship, however, clearly played a large role: twice as many Republicans as Democrats stated that the economy had gotten worse. Curiously, however, in this case the effect of partisanship was particularly visible among the better-informed voters. Why the contrast with the case concerning the deficit? The authors of the study speculate that the big difference was that President Clinton was explicitly mentioned when people were asked about the budget deficit, while the question about the economy had been phrased in neutral terms—a perfect example of the framing effect we saw earlier.

When Achen and Bartels then looked at whether and how people change their minds on important issues, they found yet more reason to be skeptical of human reason. They had access to longitudinal (across years) data concerning people's positions on abortion, which is unusual as a political issue in that opinions tend to be stable over long periods of time. They found that in the period from 1982 to 1997—when the Republican Party gradually made abortion a major issue in its platform, under the rising influence of the Christian right—people were leaving the party while retaining their previous position on abortion, women more frequently than men (which makes sense, since the issue is obviously more pressing for women). Interestingly, men who were well informed on the issue acted as the women did, while less-informed men were more likely to stay with the Republican Party through the period during which its political platform changed. This latter group was also more likely to rationalize what amounted to a change in their position on abortion in order to harmonize it with their decision to stick with the Republican Party. A similar phenomenon happened more recently when President George W. Bush pushed for privatizing social security: as it turns out, people who supported privatization did not see Bush more favorably after he announced his policy; however, people who already supported Bush did become more likely to support privatization of social security than they had been before.

Given all the disconcerting things we have learned about how the brain largely works to rationalize our views of the world—in both its pathological and standard modes—was Aristotle wrong in thinking that *the* distinctive characteristic of humanity is rationality? Not exactly. The Greek philosopher was also one of the early students of human psychology, and he was very much aware of the constant failings of the human mind. What he meant was that human beings, as far as we know, are the only animals *capable* of rational thinking, despite the fact that it doesn't come easy. That is why it is crucial to be aware of the many pitfalls of human reasoning, which we have begun to look at in this chapter and which, as we will see, affect even the quintessential application of reason to our understanding of the world: science itself (Chapter 8). Only through this awareness and constant vigilance can we hope to improve our ability to make reasonable decisions, both large and small, about everything that affects our lives. Think of it as training your brain the same way you train your muscles at the gym: both efforts achieve better results the more we take advantage of the best knowledge available about how they work.

CHAPTER 7

INTUITION VERSUS RATIONALITY, AND HOW TO BECOME REALLY GOOD AT WHAT YOU DO

> Intuition will tell the thinking mind where to look next.
> —JONAS SALK, DISCOVERER OF THE POLIO VACCINE

HAVE YOU EVER BOUGHT A CAR? HAVE YOU EVER BEEN on a date? If your answer is yes, then you may agree that often people should behave very differently in those two situations. Allow me to explain.

When you buy a car, you probably look up reviews of different models, come up with a list of desiderata in accordance with your specific needs, visit different dealerships, try out different models in person, and then sit down and ponder all this information, perhaps creating a table to help you compare your top choices. *Then* you make a choice and buy the car. At least, this is how most people *think* they ought to behave when purchasing a vehicle, even though a good number of us are likely to end up following the "gut feeling" we happen to have in response to a particular model, often for no reason other than that we look or feel good in it (and the color is definitely the right one). My point is this: we think we should act rationally when purchasing a car because it is an important investment of money and it is important for our safety. The fact that more often than not we end up following our intuition instead of our reason is just an unfortunate part—in this case—of being human.

Compare that with being on a date with someone you may get involved with. Few people would argue that the best approach is to read the other person's reviews (which in these days of electronic dating we can actually do!) or come up with lists of qualities we "must have" or "need to avoid." This is about love, right? It's not particularly romantic to invite someone to dinner and then show up with roses in one hand and a checklist and pencil (or an iPhone app) in the other. This is a situation where you are *supposed* to trust your intuition, because it's about love and commitment, not something as mundane as buying a car. Then again, one could look at the facts and argue that the decision to engage in a relationship is a hell of a lot more important—and carries a lot more consequences—than buying a car. Not to mention those oft-cited pesky statistics about the high divorce rates in modern society. Shouldn't we in fact try to keep our hormones and intuitions on hold and whip out our checklist instead?

These two examples illustrate our ambivalent attitude toward rational thinking (the checklist and library research approach) and intuition (the gut feeling, though it obviously has nothing to do with our digestive system). Indeed, we keep hearing a lot of nonsense about these two modes of thinking and about their relationship to each other. Some people think of themselves as intuitive across the board, a notion that we will see is unfounded. Others—most famously the ancient philosophers Plato and Aristotle—think that it's necessary for reason to override emotive reactions (which are often confused with intuitions). Still more people—Enlightenment philosopher David Hume among them—believe that reason is just a means to an end, but that the real drive of human action is always emotional.

In this chapter, we are going to take a look at intuition and how it relates to conscious cognitive thinking, because these are the two channels through which our minds make sense of the world. As we shall see, intuition and rational cognition can work in concert to allow us to do all sorts of interesting things—from choosing the right car to falling in love with the right person—and even to excel at whatever it is we wish to do, from playing chess to learning a musical instrument to simply being as good as we can be at our daily job.

The word *intuition* comes from the Latin *intuir*, which appropriately means "knowledge from within." Until recently, intuition, like consciousness (see Chapter 10), was the sort of thing that self-respecting scientists stayed clear of, on penalty of being accused of engaging in New Age woo-woo rather than serious science. Heck, even most philosophers—who historically had been very happy to talk about consciousness, far ahead of the rise of neurobiology—found themselves with not much to say about intuition. However, these days cognitive scientists think of intuition as a set of nonconscious cognitive and affective processes; the outcome of these processes is often difficult to articulate and is not based on deliberate thinking, but it's real and (sometimes) effective nonetheless. It was William James, the father of modern psychology, who first proposed the idea that cognition takes place in two different modes, and his insight anticipated modern so-called dual theories of cognition. Intuition works in an associative manner: it feels effortless (even though it does use a significant amount of brain power), and it's fast. Rational thinking, on the contrary, is analytical, requires effort, and is slow. Why, then, would we ever want to use a system that makes us work hard and doesn't deliver rapid results? Think of it this way: intuitions, contrary to much popular lore, are not infallible. Cognitive scientists treat them as quick first assessments of a given situation, as provisional hypotheses in need of further checking. Sometimes you have to make a decision fast and on the basis of relatively little information—so intuition is the only game in town. But if you can afford to deliberate on the issue and collect further data, then it will pay off to get your overt thinking and your unconscious thinking working together on the problem at hand.

One of the first things that modern research on intuition has clearly shown is that there is no such thing as an intuitive person tout court. Intuition is a domain-specific ability, so that people can be very intuitive about one thing (say, medical practice, or chess playing) and just as clueless as the average person about pretty much everything else. Moreover, intuitions get better with practice—especially with a lot of practice—because at bottom intuition is about the brain's ability to pick up on certain recurring patterns; the more we are exposed to a particular domain of activity the more familiar we become with the relevant patterns (medical charts, positions of chess pieces), and the more and faster our brains generate heuristic solutions to the problem we happen to be facing within that domain.

And of course, just as with everything else that we think or feel, we can now pinpoint which areas of the brain are most involved with intuition (as opposed to overt cognition). The list includes the amygdala, the basal ganglia, the nucleus accumbens, the lateral temporal cortex, and the ventromedial prefrontal cortex. The inclusion of the amygdala is particularly revealing, since, as we've seen, that's the part of the human brain that is most associated with emotions. Because of this association, intuitions are accompanied by a strong "gut feeling" that we are right. Intuitive responses and emotional responses are not exactly the same thing neurologically speaking, but they share some of the same brain circuitry and are therefore difficult to disentangle. That can be a problem when we end up feeling sure of an intuition that turns out to be wrong—we buy a lemon car, or we marry the wrong person.

The deep connection between emotions and intuition was evident in a study conducted by Christian Jordan and his colleagues at Wilfrid Laurier University. They were interested in the relationship between trust in intuition and implicit self-esteem. People can have implicit or explicit self-esteem, the former being predictive of nonverbal indicators of anxiety (skin conductance, for instance), and the latter being correlated with conscious reports of anxiety (that is, when one is aware of being anxious). Interestingly, implicit and explicit self-esteem are often uncorrelated, but Jordan and his collaborators discovered that if people trust their intuitions, their implicit self-esteem increases and implicit and explicit self-esteem become positively correlated. Conversely, if people distrust their intuitions, their implicit self-esteem goes down and the relationship between implicit and explicit self-esteem breaks down or even becomes negative.

This relationship between trust in one's intuition and implicit self-esteem is not just correlational: Jordan and his colleagues were able to experimentally manipulate their subjects' trust in intuition simply by telling half the subjects, "There is clear evidence that people who adopt an intuitive approach to decision-making are more successful in many areas of their lives," while telling the other half, "People who adopt a rational approach to decision-making are more successful." Astonishingly, this simple priming worked: when participants' trust in intuition was measured, sure enough, those who were told that intuition is a better guide for decision-making scored significantly higher in their trust in intuition than those who were told that rational decision-making is superior. Crucially, the subjects who had been experimentally manipulated to have more trust in intuition also showed augmented intrinsic and extrinsic self-esteem, with the latter two becoming positively correlated.

Why should this be? Jordan and his collaborators speculate that people experience implicit self-esteem as a particular *form* of intuition, so it follows that if they are inclined to trust intuition in general their implicit self-esteem will go up, while if they are not inclined to trust their intuition their implicit self-esteem will go down. With these findings, we can begin to appreciate how complex and subtle human behavior is, and why people are so emotionally attached to their intuitive abilities—regardless of how good their intuitions about particular problems actually are.

So intuitions are good for a number of reasons: they are effortless, they often provide us with efficient shortcuts as we tackle complex problems, and they can even alter our self-esteem through their connection with our affective responses. Still, remember the idea that intuitions should be treated as provisional hypotheses to be tested in the light of conscious reason, if at all possible (and if the effort is worth it)? That idea leads us to ask: can people learn to accord limited trust to their intuitions and to move on to full cognitive engagement when the situation requires it? Research by Adam Alter and his collaborators at Princeton, the University of Chicago, and Harvard shows us intriguing clues to the answer.

There are some standard situations in which people do switch from intuitive to explicit analyses of a problem, usually (though certainly not always) when they have something personal at stake in the outcome or when they know that they will be called to account for their decisions. However, people who are under time pressure or who are experiencing "cognitive load" (that is, they are simultaneously engaged in other tasks that deplete their brain resources) will rely on intuition and be less likely to correct errors that may arise from it. As it turns out, our brain even comes equipped with mechanisms that tell us (subconsciously) when it is or isn't time to trust our intuition—a metaintuitive mechanism about intuition itself, if you will.

Alter and his colleagues have investigated the effect on the use of intuition of what they call "disfluency," a measure of how comfortable we are with the information we are receiving. It turns out that the more disfluent we are about something, the less we rely on intuition and the more we engage in full-bore analytical reasoning. Neurologically speaking, disfluency triggers the anterior cingulate cortex, which activates the prefrontal cortex, where much of our analytical thinking takes place.

What causes disfluency? A number of factors are involved, some of them very simple and conveniently easy to manipulate experimentally. For instance, Alter and his group simply provided information about a particular problem or situation to their subjects in one of two forms: written in a clearly legible type font or written in a type font that was a bit more difficult to read. (Follow-up experiments eliminated the possibility that the crucial difference between the two treatments was a simple slowing down caused by the less legible font.) Subjects who received the disfluent write-up provided more accurate responses to the test as a direct result of their more systematic (less intuitive) processing of the information thus made available. It seems that the brain really needs for things to get difficult before it can be bothered to engage its more sophisticated, effortful, and time-consuming thinking apparatus!

The experiments conducted by Alter's group even led to the discovery that self-induced manipulations can do the trick. Subjects were asked to furrow their brows in an effort to simulate the expression of someone engaged in deep thinking, while the control group was asked to puff out their cheeks, presumably an activity unrelated to the type of thinking one does. The people who furrowed their brows turned out to be less confident in their intuitions, as a result of which they engaged in more effortful thinking, which led to better results on their tests. The next time you want to trick your brain into behaving intellectually, just imitate some of the stereotyped postures of an intellectual and your prefrontal cortex will get the message.

To achieve the most productive balance between intuition and analytical thinking is obviously crucial to making the best decisions we can in our everyday lives, but it turns out that this balance is also important for businesses (and governments). Accordingly, some authors have begun to pay attention to how managers in the business world achieve this balance. The prevalent business culture until recently has put an emphasis on deliberation and on the explicit use of all the available information to achieve the best possible results. But as Marta Sinclair and Neal Ashkanasy point out in a paper they coauthored on this subject, analytical decision-making in business has an efficiency rate of about 50 percent. Although one could very well argue that a 50 percent success rate when dealing with complex problems characterized by dozens of variables is actually not bad at all, Sinclair and Ashkanasy and other researchers began to investigate whether an increased emphasis on intuition in business decisions has an effect on this efficiency rate. Their conclusion? That the best approach to business decisions is an integrative model in which domain-specific (that is, based on expertise) intuition and rational thinking are used in concert, just as cognitive scientists have suggested we do in other areas of life. It seems that the old counterposition between intuition and analysis is finally giving away to a more—shall we say?—reasonable understanding of the human mind and how to get the most out of it.

Throughout this discussion so far you may have noticed that I have made no mention of gender or culture, and you may be eager to ask: Isn't it well known that women are more intuitive than men? And that Asian cultures emphasize more holistic thinking as opposed to the Western analytical approach? It turns out that there is good evidence for the latter generalization (though we do not know exactly why), and very little, if any, support for the former. The first hint that we should be suspicious of the idea that women are more intuitive than men comes from our earlier realization that there simply isn't any such thing as general purpose intuition. We can be very intuitive at X, and horribly not intuitive at anything but X. And sure enough, neither the research conducted by Jordan and his collaborators nor the study by Sinclair and Ashkanasy turned up much evidence of gender-specific differences in intuitive abilities. Indeed, the whole—perennially popular—cottage industry purveying "men are from Mars and women are from Venus" nonsense has been thoroughly and convincingly debunked by a number of authors, including Cordelia Fine (in her *Delusions of Gender*) and Rebecca M. Jordan-Young (author of *Brain Storm: The Flaws in the Science of Sex Differences*), though I'm sure the myth will persist, as myths are prone to do.

What about intercultural differences? Those are easier to verify and quantify, though we can only speculate about why these differences are present. For instance, research conducted by Emma Buchtel and Ara Norenzayan showed that Korean college students consistently think intuition is more important than logic, while American students rank the two approaches in reverse. (Although a look at the details reveals that the first difference was statistically significant while the second one was not.) The same researchers also compared Canadians of European origin with Canadians of East Asian origin: they found that both groups think of intuitive people as more social, but East Asians (not Europeans) also think of them as wiser and more reasonable.

This kind of study is, of course, entirely descriptive. It is neither prescriptive (it doesn't tell us which approach is more effective) nor causally explicative (it doesn't tell us why the intercultural differences are there to begin with). From what we have seen so far, it seems likely that, from a prescriptive perspective, Westerners would do well to rely a bit more on intuition, while Asians would benefit from a more systematic use of analytical thinking. And there are two major types of explanations for why different cultures seem to have these propensities, though it is hard to imagine how either one could be tested empirically. (Moreover, they are certainly not mutually exclusive and could very well reinforce each other.)

One possibility is that Asian cultures are characterized by more social interconnectivity and mutual interdependence than Western ones and that holistic-intuitive thinking simply works better in the Asian social environment, while atomistic-analytic thinking works better in the Western one. The other explanation is historical: Western thought arose out of the philosophy of ancient Greece, the birthplace of logic and analysis, while Asian thinking has been influenced by Confucian and Taoist traditions—such as the concept of *wu-wei*, or effortless action, which can easily be interpreted as a form of intuition. Whether certain historical roots gave birth to a particular type of culture or certain historical traditions took hold *because* Aristotle and Confucius, for instance, found themselves in different environmental milieus, it is hard to tell. If the latter (which I do think is more likely), we are still left with no explanation for why some human cultures tend to be more socially interdependent while others favor individualism. The difference is highly unlikely to be genetic in nature (there are very few systematic differences in the genetic makeup of different ethnic groups), so perhaps the answer can be found in differences in the physical environment combined with historical contingencies.

Be that as it may, the different emphases that Asians and Westerners put on intuitive versus analytical approaches should be taken into consideration by both businesses and governments when dealing with each other. As Buchtel and Norenzayan note, for instance, educators from the two traditions sometimes disparage each other: some Asian educators think that Western students are dogmatic and simplistic in their approach to problems, and some Western educators think of Asian students as not sufficiently rigorous from a logical perspective. This "mutual yuck" effect, as Buchtel and

Norenzayan call it, ends up affecting international trade and intergovernmental relations, as when American representatives to the World Trade Organization complained that their Chinese counterparts were neither explaining nor substantiating their decisions; to the Chinese, this was an odd complaint based on a refusal to see problems and solutions holistically. Learning about how the brain works apparently can affect not only our individual lives but also human interactions in the wider world.

There is another aspect to the question of intuition versus conscious thinking that affects our quality of life, and that has to do with research showing how people get better at what they do or get stuck in it. If you are lucky enough to have a job you really enjoy, the kind that makes you look forward to getting up on Monday mornings, you are also likely to already be in a rarefied group of human beings who wish to be better for the sake of being better at their chosen profession—a goal that significantly affects how meaningful you think your life is. But regardless of whether you do your job because you love it or simply because it's a good way to make sure you can pay your bills and provide for your family, doing it better is a ticket to more satisfaction and possibly a raise or promotion. Or consider another possibility: maybe your job is just a job, and that is why you love to spend your free time playing an instrument, practicing a sport, or engaging in another activity that enriches your life. Even then, the better you are at your hobby the more satisfaction you will gain from it. In any of these cases, you may want to know about research on expertise.

An "expert" is someone who performs at a very high level in a given field, be it medicine, law, science, chess, tennis, or soccer. As it turns out, people become experts (or simply, much much better) at what they do when they use their intuition and conscious thinking in particular ways. Research on acquiring skills shows that, roughly speaking, and pretty much independently of whether we are talking about a physical activity or an intellectual one, people tend to go through three phases while they improve their performance. During the first phase, the beginner focuses her attention simply on understanding what it is that the task requires and on not making mistakes. In phase two, such conscious attention to the basics of the task is no longer needed, and the individual performs quasi-automatically and with reasonable proficiency. Then comes the difficult part. Most people get stuck in phase two: they can do whatever it is they set out to do decently, but stop short of the level of accomplishment that provides the self-gratification that makes one's outlook significantly more positive or purchases the external validation that results in raises and promotions.

Phase three often remains elusive because while the initial improvement was aided by switching control from conscious thought to intuition—as the task became automatic and faster—further improvement requires mindful attention to the areas where mistakes are still being made and intense focus to correct them. Referred to as "deliberate practice," this phase is quite distinct from mindless or playful practice. Think of a soccer (or football or baseball) player, for instance. Just mindlessly kicking the ball back and forth or playing a no-stakes pickup game isn't going to improve her skills, no matter how many hours a week she does it, because those activities simply reinforce the automatic, intuitive style of play she has acquired up to that point. To make it to the next level, she has to concentrate on the plays and moves that still don't come easily or naturally for her, and in order to do that she has to identify those problem areas (probably with the help of a coach) and then focus her mind on overriding her intuitive and already ingrained way of handling those very areas. It's hard work, and it requires a fine balance between intuitive (fast, automatic) and conscious (slow, mindful) thinking. Without it, however, the player's skill development becomes "arrested," stuck at an intermediate level that is likely to become increasingly frustrating and to affect not just her career (or hobby, if she is not a professional athlete) but her quality of life (if the activity constitutes an important area of her life).

Researchers have also figured out how long it takes to develop simple proficiency in a given field or activity versus actual expertise. And again, roughly speaking, the results are about the same regardless of whether we are talking about playing the violin, chess, or tennis. The good news is that simple proficiency can be achieved in a matter of weeks or months. The bad news is that expert-level proficiency requires on average ten years of practice! Chess players don't get good enough to compete in international tournaments if they don't engage the game for about a decade (and it takes *another* decade to reach the level of chess master, if one is capable of that much). Moreover, the ten-year (approximate) rule applies even to particularly talented individuals, like a child prodigy.

Why should this be? There are a variety of reasons, but two are especially important: one needs to develop the ability to *anticipate* problems, and this in turn is often the result not just of knowledge of a given field but of *structured* knowledge. For instance, studies of tennis players show that the best ones don't simply react to whatever their opponent is serving them but are capable of anticipating where the ball will go before the opponent even hits it. There is no magic to this ability, of course; it is an acquired intuitive skill made possible by the brain having seen enough similar situations to extract patterns and thus predict where the ball is most likely to go from the anticipated angle of impact on the opponent's racquet. Similarly, what makes a chess player good is not that he has memorized a high number of specific configurations on a chessboard and is able to recall them at will; human chess players are not computers—they don't store information that way. Rather, the chess master's long experience has allowed him to acquire structured knowledge, a built-in understanding of the strategy of the game that makes it possible to intuit the likeliest ways out of any difficult configuration on the chessboard. Indeed, when chess masters are presented with random board configurations—that is, configurations that are not likely to arise in an actual game—the fact that they can't recall them with any particular accuracy shows that their memory is of structured situations, not of the simple layout of pieces on the board.

Cindy Hmelo-Silver and Merav Green Pfeffer have investigated this difference between superficial and structural knowledge in the particular case of people's understanding of something as mundane as aquaria. They compared how four groups of people understand aquaria: children, "naive" adults (adults with no particular interest in the subject matter), and two types of experts—biologists with an interest in ecology and hobbyists who like, build, and care for aquaria. Not surprisingly, children and naive adults displayed a much simpler understanding of the workings of an aquarium, often resorting to one type of causal explanation and failing to appreciate the intricacies of the system. Experts, on the other hand, were appreciative of the systemic functioning of an aquarium and could describe multiple causal pathways affecting the enclosed ecosystem. But here is the interesting part of Hmelo-Silver and Green Pfeffer's findings: the two groups of experts also differed dramatically in the kind of knowledge of aquaria they had built. Regarding aquaria as microcosms of natural ecosystems, biologists explained them at an abstract-theoretical level in terms of the science of ecosystems. Hobbyists, on the other hand, built their mental model around the practical issues associated with filtering systems, feeding systems, and generally anything that played a direct role in keeping the aquarium looking good and the fish healthy. Not only is there a difference between naive and expert knowledge, but there is more than one way to acquire expert knowledge, guided not just by the intrinsic properties of the system but also by the particular kinds of interest that different individuals have in that system.

But, one might object, all this talk of improving one's skills, of becoming chess masters and expert soccer players, surely neglects the idea of talent. Some people are just innately good at certain things, and others will never get to Carnegie Hall, no matter how much they practice. Perhaps, but hard evidence for the idea of innate talent is surprisingly hard to find, as explained in detail in David Shenk's *The Genius in All of Us: Why Everything You've Been Told About Genetics, Talent, and IQ Is Wrong*. As Philip Ross explained in an article in *Scientific American*, very often people who talk about talent confuse innate ability with precocity. The two are certainly not the same thing: some children may display an early aptitude for, say, music, but from then on it is practice, practice, practice that turns them into actual prodigies (and gets them to Carnegie Hall). Having studied nature-nurture issues for my entire career as a scientist, I certainly am not about to downplay the importance of one's genetic makeup and its effect on all aspects of our lives. But it is unfortunate that in many people's minds the step from acknowledgment of genetic differences to belief in some sort of entirely unsubstantiated and downright pernicious form of genetic determinism seems to be extremely short; indeed, an oversimplified belief in innateness probably accounts for a substantial degree of human unhappiness, including persistent "scientific" attempts to show that particular ethnic groups (or women) are innately inferior at X whenever X happens to be something that furthers the interests of members of another ethnic group (or of males).

There is one more issue that we need to be aware of when it comes to expertise, since it affects so many aspects of our lives. There are, demonstrably, fields where alleged experts are no such thing at all, and you will pay with money, time, and emotional resources if you fall for their claims. Anders Ericsson, in *The Cambridge Handbook of Expertise and Expert Performance*, refers to studies that show, for instance, that so-called wine experts perform only slightly better than regular wine drinkers when they cannot read the label of the wine they are tasting. Knowing this could save you hundreds of bucks at the liquor store, and of course, if more widely appreciated, could shake the foundations of the multimillion-dollar wine industry. Similarly, Philip Ross points out that the evidence shows that psychiatric therapists with PhDs don't actually help their patients much more than those with a master's degree, and that—more ominously considering the worldwide financial upheavals of recent years—professional stockbrokers don't do any better than amateurs at picking winning stocks. But I'm sure you knew that already.

CHAPTER 8

THE LIMITS OF SCIENCE

> There is something fascinating about science. One gets such wholesale returns of conjecture out of such a trifling investment of fact.
>
> —MARK TWAIN

'M NOT A DOCTOR, BUT I PLAY ONE ON TV!" "NINE OUT of ten doctors recommend. . . . " These are common phrases that remind us of just how much science is revered by the general public—with good reason, but also with some glaring exceptions. There are excellent reasons to hold science in high regard, from its fundamental discoveries about the origin of humanity and the origin of the cosmos to the countless technological and medical benefits that have originated from scientific research. Most people do feel—if asked in the abstract—that scientists as a group deserve the high reputation that their profession has achieved in modern society. Moreover, some scientists have even become household names and appear in societal discourse outside the realm of science, from Albert Einstein to Stephen Hawking, Charles Darwin, and Richard Dawkins.

When it comes to the specifics, however, a large portion of the public seems to ignore or even be downright hostile to a number of scientific ideas. About 30 percent of Americans believe in astrology, a notion that was debunked centuries ago; about 40 percent still believe in creationism, an idea that has been rejected by science for more than a century and a half; in a recent count (a Gallup poll in March 2010) only slightly more than 50 percent of Americans agreed that climate change is real and largely caused by humans, even though the consensus within the scientific community in that regard has been overwhelming for years; and almost 20 percent of the population believe that vaccines cause autism, even though the only paper claiming such a connection has been shown to be a fraud and dozens of follow-up studies have revealed no link whatsoever.

The idea underlying this chapter is that science is neither the new god nor something that should be cavalierly dismissed. As a society, we need a thoughtful appreciation not only of how science works but also of its power and its limits. This isn't just an (interesting, I would submit) intellectual exercise: how we think about science has huge personal and societal consequences, affecting our decisions about everything from whether to vaccinate our children to whether to vote for a politician who wants to enact policies to curb climate change. We cannot all become experts, especially in the many highly technical fields of modern science, but it is crucial for our own well-being that we understand the elements of how science works (and occasionally fails to), that we become informed skeptics about the claims that are made on behalf of science, and that we also do our part to nudge society away from an increasingly dangerous epistemic relativism. Jenny McCarthy, the celebrity who has been pushing for people to stop vaccinating their children because of the threat of autism, famously said, "My science is Evan [her son]. He's at home. That's my science." As much as this kind of declaration elicits empathy for a distressed mother, the fact is that, no, science is not and cannot be done through the personal experience of a single individual with no technical training, just as nobody other than a highly trained brain surgeon, working in a properly equipped facility, should attempt brain surgery.

The first thing to understand about science is that scientific reasoning is a refined form of the two basic types of reasoning that we all use; in this (limited) sense, science truly is just common sense writ large. The two types of reasoning in question are deduction and induction. Let's take a look. Deductive reasoning takes this form:

Premise 1: All philosophers like arguing.
Premise 2: Massimo is a philosopher.
Conclusion: Therefore, Massimo likes arguing.

Deductive reasoning is "truth-preserving," as philosophers like to say. That is, if the structure of the argument is valid (as is the one above, which can be generalized as: if P, then Q; P; therefore Q, where P and Q stand for any meaningful statement), and if the two premises are true, then the conclusion must be true. So there are basically two ways in which a deductive argument can go wrong: when its structure is flawed (the argument is invalid) or when one or more of its premises are not true (the argument is unsound).

Here is an example of invalid deduction:

Premise 1: If it snows, the pavement becomes wet.
Premise 2: The pavement is wet.
Conclusion: Therefore, it snowed.

Can you catch the problem? It helps to see the formal structure of the argument: if P, then Q; Q; therefore P. Notice the similarity to the first example, which may make it temporarily difficult to see what is wrong with this second one. But think about it for a minute: clearly there may be many other reasons why the pavement is wet—for instance, because somebody used a powerful water jet to clean it up. (If you live in New York City you'll know instantly what I mean.) Indeed, this kind of mistake is so common that it has its own name: it's called *the fallacy of affirming the consequent* (the consequent is Q). So when you look at a deduction, the first thing you want to check is whether the argument being presented is valid—that is, whether it is structurally correct.

What about the first example, which is in fact valid? Is the conclusion true? The answer is yes in this particular case: as it turns out, I do like arguing (in a friendly manner, my favorite motto being "Truth springs from argument amongst friends," by David Hume). But the conclusion wouldn't necessarily hold for another philosopher. Why? Because while the second premise is true (I am indeed a philosopher), the first isn't: not *all* philosophers like arguing. I know that from personal experience, and I'm sure that statistics could be brought to bear on the issue. So the argument is valid (structurally correct) but unsound, because at least one of its premises is incorrect.

Why is it interesting to know any of this as far as our own lives are concerned? Because deductive reasoning is at the foundations of logic and mathematics, both of which are much more rigorous than science. So, if we can already find issues with deduction (and hence with logic and math!), things are bound to get even more intriguing when we move to the second fundamental type of reasoning that constitutes the bread and butter of science: induction. Philosopher Francis

Bacon (1561–1626) was one of the early modern theorists of the scientific method, and he thought that its very foundation was provided by inductive reasoning. There are several types of induction, but for the purposes of our discussion let's simply think of it as that general kind of reasoning that allows us to infer things that we don't know from the things that we do know. For instance, I can reasonably infer that the sun will rise tomorrow, even in the absence of any technical knowledge in astronomy, because we have countless observations of past instances in which the sun has risen and there is no reason to believe that whatever mechanism has made that possible in the past will not hold also tomorrow. In other words, induction works because we extrapolate past experience to future happenings, and we are justified in doing so as long as the mechanisms (or laws of nature) that have worked in the past continue to do so in the future. This process of inductive inference based on empirical evidence is largely what it means to do science.

Predictably, however, there is a problem with induction, and it is a big one. It was first pointed out by David Hume in the eighteenth century and best exemplified by a thought experiment proposed by Bertrand Russell (1872–1970) two centuries later. Russell asked us to imagine an inductivist turkey (the original story actually featured a chicken) who is brought to a new farm one day and begins to take notes (that is, collects empirical data) on what happens to him. After a few days, he realizes that he is being fed every morning around 7:00. Being a rigorous inductivist, however, the turkey is wary of drawing inferences about the future on the basis of few data, so he keeps accumulating observations and refrains from making predictions. Eventually, after 364 days of data gathering, the inductivist turkey finally feels confident of his knowledge base and hazards a prediction: he will be fed tomorrow morning at 7:00. Alas, that morning happens to be Thanksgiving, and the turkey is instead "prepared" to be served for dinner.

The sad story of the inductivist turkey illustrates one of the major problems with inductive reasoning: it is not truth-preserving (unlike deduction, when properly carried out). Hume's critique, however, went even further. He pointed out that the only reason we think that induction is a good way to proceed about making inferences concerning the world is because it has worked in the past. This observation may seem innocuous enough until we realize, upon a moment's reflection, that Hume was saying that our endorsement of induction is in itself a form of induction: we argue that induction works because it has worked in the past, thereby applying inductive reasoning to justify induction. This, Hume observed, is an instance of circular reasoning, one of the most elementary logical fallacies. Oops! If this doesn't immediately bother you, think about it for a while and you'll find that it does. Let me put it clearly: Hume's critique amounts to saying that there is no rigorously logical foundation for the entire enterprise of science! Add to that the failure during the early part of the twentieth century (by a number of people, including Bertrand Russell himself) to find a logically tight foundation for math, and all of a sudden we are looking at some pretty solid reasons to doubt the truth of both math and science (and, incidentally, logic itself).

Naturally, philosophers simply won't stand for something like this, and plenty have tried to come to the rescue of science. (Scientists themselves aren't usually bothered too much by this sort of issue, though arguably they should be.) A valiant attempt was mounted by Karl Popper (1902–1994), who thought he figured out a way to enlist deduction to solve the problem of induction. His attempt ultimately failed, but it has much to teach us about the nature of both science and philosophy, so let us take a closer look.

Popper thought that what scientists really should be doing was trying not to prove but to disprove their theories, a process he called "falsification." The idea was that—partly because of the problem of induction—one can never prove a proposition, no matter how many new facts agree with it, since new evidence may come up later that disproves it. But, Popper thought, once a theory has been falsified, that's the end of the story—it will never again emerge to fight another day. In other words, science makes progress because scientists learn what doesn't work and discard it, thereby getting closer and closer to the truth.

There is quite a bit that is appealing about this idea, not the least of which is that it is based on the application of deductive logic. To see how, consider the following deductive argument:

Premise 1: If Newtonian mechanics is true, then light should bend around massive objects by a certain amount.
Premise 2: Light bends around massive objects by a larger amount than predicted by Newton.
Conclusion: Newtonian mechanics is wrong.

This argument is both valid and sound. It is valid because its form (if P, then Q; not Q; therefore not P) is correct, and it is sound because as it turns out astronomers at the beginning of the twentieth century did discover that light is bent by massive objects (like the sun) to a degree that is not compatible with Newton's predictions. Consequently, they permanently discarded Newtonian mechanics in favor of Einstein's general relativity (which predicted the correct amount of bending). Problem solved, then—science works by the process of falsification, and Popper incidentally also solved the problem of induction.

Well, not quite. The history of science provides us with plenty of examples of scientists simply not behaving the way Popper said they should, and for good reasons. For instance, in 1821 the astronomer Alexis Bouvard had calculated a series of tables predicting the position of what was then thought to be the outermost planet in the solar system, Uranus. The problem, as Bouvard soon recognized, was that there was a significant discrepancy between the predictions and the actual positions of the planet in the sky. According to a strict interpretation of falsificationism, Bouvard and his colleagues at that point should have rejected Newton's theory, as it was manifestly and systematically incompatible with a large set of data. But they didn't. Instead, Bouvard immediately intuited the obvious answer: there must be another planet influencing Uranus's orbit, thus accounting for the anomaly. A few years later, on September 23, 1846, Neptune was discovered within 1 degree from the position calculated by the astronomer Urbain Le Verrier. Newtonian theory was safe (for the moment), and the solar system had acquired a new member!

This episode illustrates that the actual practice of science is very different from what Popper at first proposed, and in particular that scientists do not throw out a hypothesis for which there is a lot of confirmatory evidence, even in the face of some disconfirming evidence, until they absolutely have to, and probably not until they have a better alternative handy. (Newtonian physics survived until the advent of relativity.) A better way to think about how science works was proposed by Thomas Kuhn (1922–1996), partly in response to Popper's ideas. (This, incidentally, is how philosophy makes progress: people analyze other people's ideas and show the logical flaws in them, until the original idea is either modified and improved or completely abandoned.)

Kuhn suggested that science works in two modes, one that accounts for most of everyday scientific activities, and another that explains when new theories are proposed and old ones abandoned. For Kuhn, everyday science is about "puzzle solving"—scientists applying a particular conceptual framework, which Kuhn called a "paradigm," to the solution of specific questions. So, for instance, an astronomer applies the Newtonian paradigm to the calculation of the orbits of the known planets (pre-Neptune), a biologist uses the Darwinian paradigm to investigate the mating habits of a particular species of butterflies, and a geologist deploys the continental drift paradigm to explain the geographical distribution of a given group of fossils.

Most of the time this puzzle solving goes on quite well and amounts to what Kuhn called "normal science." However, from time to time some of these puzzles are not solvable within the reigning paradigm. This does not usually bother scientists, who simply move to the next puzzle in order to continue their careers. But occasionally the growing number of unsolved puzzles begins to cause a stir in the scientific community. People start tinkering with the paradigm itself, and if this doesn't work, the science in question enters a period of crisis, from which it emerges only when a new paradigm has been identified and accepted by the relevant scientific community. This is precisely what happened at the beginning of the twentieth century in physics, when a mounting number of problems with Newtonian mechanics and classical physics gave rise to Einstein's relativity and quantum mechanics.

Even Kuhn, however, isn't the end of the story: successive philosophers of science have found problems with his account of science and come up with interesting alternatives in turn. But this isn't a book about the philosophy of science; it is about appreciating the intricacies of both science and philosophy so that we can be better informed and more sophisticated thinkers about life in general. So we will move on to one last debate concerning the nature of science before trying to figure out if we can trust science at all, given all these problems! (The answer will turn out to be yes, but in a somewhat cautionary and provisional manner.)

The debate in question is between so-called realists and antirealists about scientific theories. Let's first clear up one possible source of confusion: antirealists are not people who claim that science is a fantasy, or that scientific theories are arbitrary and have no connection with reality. Indeed, the word realism here is probably a bad choice, but sometimes labels stick and it's more straightforward to use the terms adopted by the community of scholars who study these things. So, a realist about scientific theory is someone who says that theories in science describe reality as it actually is, or as close to it as human reasoning and observational powers can get us. Most scientists, needless to say, are realists, and a good number of philosophers of science are too. When physicists who are realists talk about electrons, for instance, they are not conjuring a hypothetical construct to help make sense of the data, but rather are referring to physical objects out there with the characteristics of electrons—even though we cannot actually observe them. (The latter bit is important, since much of the discussion hinges on the status of the "unobservables" in science—that is, the theoretical entities that are necessary for a theory to work but cannot be directly observed.)

Now, one might think, how can anyone seriously doubt that electrons exist? Isn't it because of the existence of electrons that when we turn a switch connected to an electrical circuit the lights in our apartment turn on? Well, that's the realist explanation of what is happening. But the antirealist will quickly point out that plenty of times in the past scientists have posited the existence of unobservables that were apparently necessary to explain a given phenomenon, only to discover later on that such unobservables did not in fact exist. A classic case is the aether, a substance that was supposed by nineteenth-century physicists to permeate space and make it possible for electromagnetic radiation (like light) to propagate. It was Einstein's theory of special relativity, proposed in 1905, that did away with the necessity for aether, and the concept has been relegated to the dustbin of scientific history ever since. The antirealists will relish pointing out that modern physics also features a number of similarly unobservable entities, from quantum mechanical "foam" to dark energy, and that the current crop of physicists seems just as confident about the latter two as their nineteenth-century counterparts were about aether.

The most compelling argument on the side of the antirealists is the so-called underdetermination of theory by the data. It is easy to show that any particular set of empirical observations—the basic point of reference for any scientific theory—is compatible with literally an infinite number of different theories, a good number of which will not be trivial variations on a single theory. Want an example? The rage in theoretical physics for the past three decades (and counting) has been something called "superstring theory." It is supposed to conceptually unify the two dominant subtheories in physics, general relativity and quantum mechanics (which make different predictions when applied to some of the same phenomena, thus suggesting that there is something wrong or incomplete about them). Superstring theory at the moment cannot be tested experimentally because it does not make any new verifiable prediction that is not also made by preexisting theories like relativity and quantum mechanics. That's bad enough, since the hallmark of a scientific theory is supposed to be its empirical verifiability. But the really bad news is that superstring theory is not actually a theory—it is a *family* of mathematically related theories, estimated to number about 10^{500} members. That's an astronomically high number of theories (to write it out, you'd have to write a 1 followed by a whopping five hundred 0s!), and there simply is no way that we will ever have sufficient experimental data to discriminate among such a huge number of theories, which means that any (or none) of them could be true and we will simply not be able to tell. The data, no matter how much of it there is, will massively underdetermine—that is, won't be able to pick out—the correct theory.

Things begin to look grim for the realists now, don't they? Nonetheless, they do have a pretty convincing countermove, which is amusingly referred to as the "no miracles" argument. It goes like this. Although it may be true that the available data never pick the absolute best theory and only that theory, would it not also be very odd indeed, if not little short of a miracle, if scientific theories had nothing to do with the way the world really is? What are the odds—argues the realist—that a complex theory like Einstein's general relativity, or quantum mechanics, would be able to predict an astounding number of facts about the world, and to an astounding degree of precision, if it was not in some meaningful sense *really* describing the way the world is?

And yet, the antirealist would argue in turn, this is precisely what has happened even in the recent past (an observation colorfully known as the "pessimistic meta-induction"). Newtonian mechanics was considered the dominant theory in physics for centuries, and it is still used today to calculate the trajectories of projectiles, such as space probes. And it works. But we also know that it is profoundly wrong. Although it is sometimes said that Newtonian mechanics can be derived as a special case via mathematical approximation from the general theory of relativity, it is also true that the underlying concepts of space-time proposed by Newton and Einstein are qualitatively different; at best, only one can be correct, certainly not both. Even so, the "no miracles" argument does have some force, and the debate between realists and antirealists, once we pass the simple level sketched earlier, quickly becomes very technical, and it remains unresolved to specialists in the field to this day.

There is, however, a third way to look at science, a way that acknowledges both of the main points that I have tried to explore throughout this chapter. On the one hand, it is undeniable that science works; scientific discoveries do make a practical difference in our lives, and we do understand the universe better because of science. The theory of evolution, quantum mechanics, and the idea of continental drift are not "just" theories—they are sophisticated ways to comprehend how the world works. Cars, computers, airplanes, and space probes, not to mention the human genome project or the various medical treatments we use to improve our well-being, don't work because of magic, and they are not a matter of opinion. On the other hand, it is also true that most scientific theories have at some point or another been proven to be wrong, which means that we don't have any reason to believe that the current ones will not also join the dustbin of history. Indeed, even the most spectacular and well-regarded theory in contemporary science—the standard model arising from the conjunction of general relativity and quantum mechanics—is already known to be wrong in some fundamental respects, which is why physicists keep looking for a broader, more unifying theory of the deepest foundations of reality.

How, then, do we resolve this seeming contradiction between science being in some sense undeniably wrong and in another sense equally, undeniably right? Through an idea called "perspectivism." To understand how this works, consider a simple example first: color perception. In some sense, colors are the result of objective facts about the world: a certain electromagnetic radiation, characterized by a given wavelength, hits a particular material at a given angle; the light so reflected is captured by our retina and excites certain pigments inside our eyes, which in turn activates certain neural pathways, with the result—after a surprisingly sophisticated further processing by the brain—that we see, say, "red." In another sense, however, the way I perceive color is irreducibly subjective: it boils down to a first-person experience that simply cannot be shared by anyone else in the exact same way. Besides, you and I may be looking at precisely the same object and yet judge it to be of a different color, for a variety of reasons: because we are looking at it from different angles, or at different times and therefore in different light conditions, or maybe because I'm partially color-blind (I am) and you are not. In other words, even though there are objective facts about the world that affect our perception of color, there is also significant room for different *perspectives* about that aspect of the world. The physical facts pertinent to electromagnetic radiation and material properties delimit what can and cannot be perceived by a certain biological system of pigments and neurons, but the complexity and variability of the latter leaves room for a number of subjective interpretations of what is going on.

Scientific perspectivism applies the same idea to the process of science: scientific knowledge is both objective and subjective, because it results from a particular perspective (the human one) interacting with how the world actually is. The result is that our scientific theories will always be tentative and to some extent wrong (as the antirealist maintains), but will also capture to a smaller or greater extent some important aspect of how the world actually is (as the realist maintains). Science provides us with a perspective on the world, not with a God's-eye view of things. It gives us an irreducibly human, and therefore to some extent subjective—yet certainly not arbitrary—view of the universe.

Now, why should any of this be of concern to the intelligent person interested in improving her or his well-being through the use of reason? Because a better understanding of how science actually works puts us in the position of the sophisticated skeptic, who is neither a person who rejects science as a matter of anti-intellectual attitude nor a person who accepts the pronouncements of scientists at face value, as if they were modern oracles whose opinions should never be questioned. Too often public debates about the sort of science that affects us all (climate change, vaccines and autism, and so on) are framed in terms of alleged conspiracies on the part of the scientific community on one side and of expert opinion beyond the reach of most people on the other side. Scientists are just like any other technical practitioners and in very fundamental ways are no different from car mechanics or brain surgeons. If your problem is that your car isn't running properly, you go to a mechanic. If there is something wrong with your brain, you go to the neurosurgeon. If you want to find out about evolution, climate change, or the safety of vaccines, your best bet is to ask the relevant community of scientists.

Just as with car mechanics and brain surgeons, however, you will not necessarily find unanimity of opinion in this community, and sometimes you may want to seek a second or even a third opinion. Some of the practitioners will not be entirely honest (though this is pretty rare across the three categories I am considering), and you may need to inquire into their motives. Scientists are not objective, godlike entities, dispensing certain knowledge. They have a human perspective on things, including the field in which they are experts. But other things being equal, your best bet—particularly when the stakes are high—is to go with the expert consensus, and if a consensus is lacking, you're better off going with the opinion of the majority of experts. Keeping in mind, of course, that they might, just might, be entirely wrong.

There is one area of scientific inquiry that is both very new and of particular interest to our quest for how to live an intelligent life: cognitive neuroscience. You've seen the colorful brain scans, and you've probably heard the claims that studying the brain makes sense of just about everything human. "My brain made me do it" is likely to surface as a defense (admittedly a very strange one) in court cases, and you may find yourself serving on a jury that has to make sense of the bold new science of human behaviors and motivations. In Part III, we will look at what this approach has to say about our will to change our lives, the origin of our decisions, our propensity to fall in love, and even the way we handle friendships. As usual, we'll add a bit of philosophical reflection to the mix to get beyond the empirical facts as we assess what they mean for our lives and how to improve them.

PART III

WHO AM I?

CHAPTER 9

THE (LIMITED) POWER OF THE WILL

> I generally avoid temptation unless I can't resist it.
> —Mae West

IF YOU HAVE A PROBLEM QUITTING SMOKING, THERE IS A simple solution, at least potentially. Your neurobiologist could soon be able to inhibit your insula—a small portion of the cerebral cortex in each of your brain's hemispheres—and voilà, your addiction will be gone. Unfortunately, there is no free lunch to be had in neurobiology, so the operation would probably have some unpleasant side effects: you would experience loss of libido, become apathetic, lose your capacity to emotionally appreciate music, and develop a peculiar inability to distinguish fresh from rotten food. But at least your chances of getting lung cancer will go down dramatically! Welcome to the strange world of human volition, an ancient topic of discussion among philosophers and an active field of research in cognitive science.

The ability to make willful decisions—to exercise volition—is fundamental to the meaning of being human. With our need to feel in at least partial control of our lives, we must have the freedom to make choices between different courses of action. Our entire system of justice (Chapters 14 and 15) is based on the idea of moral responsibility, which in turn hinges on the possibility that we do make free choices. If we can't, if all our actions are determined by forces outside of our control, then it makes no sense to talk about morality. Further, without a sense of ownership of our actions, we couldn't even meaningfully take credit for what we do, no more than a computer programmed to play chess can take credit for beating a human chess master. The computer just did what it was programmed to do—no more, no less.

Not only is the concept of free will (as philosophers put it), or volition (a term more frequently adopted by cognitive scientists), fundamental to our conception of ourselves and of others, but the concept is also related to the idea of willpower. Our ability to make hard choices and stick with them is another aspect of the human condition that generates admiration for those of us who seem to display a lot of it, as well as criticism of those of us who appear to be deficient in it. Just think of the roots of the Christian idea of sin, which originated in the lack of willpower displayed by our paradise-dwelling ancestors who fell for a simple temptation offered by Lucifer.

Even today willpower features prominently in social and popular discourse and is sometimes elevated to an almost mystical level. Not long before I wrote the first draft of this chapter, US Representative Gabrielle Giffords of Arizona was shot in the head at close range by a lunatic who also fired on eighteen other people, killing six. During her recovery, the newspapers were full of quotes to the effect that it was her fierce "spirit" that was helping her throughout—even though the research clearly shows that willpower is, well, powerless when it comes to major health issues. For instance, Naoki Nakaya and his colleagues at the Institute of Cancer Epidemiology of the Danish Cancer Society conducted a large study of 60,000 subjects and followed them for thirty years; they found absolutely no connection between personality traits and the likelihood of surviving cancer. And for any positive anecdotes like the ones involving Representative Giffords, we can easily find matching stories of people who fought just as hard but lost a similar battle.

There is a dark side to this idea that we can overcome all sorts of difficulties if we just try hard enough, and that is the pernicious consequence that if we fail it must in some important sense be our own fault. This is, for instance, the sort of callous nonsense that is propagated by books like *The Secret* by Rhonda Byrne, which has become popular in part through the complicity (surely in good faith) of celebrities like Oprah Winfrey. The basic idea of *The Secret* is that through a "law of attraction" positive thoughts bring about positive outcomes, while negative thoughts bring about negative outcomes. Needless to say, there is no such thing as a law of attraction—this is New Age metaphysical baloney. But it is easy to turn the idea around and conclude that since a negative thing happened to someone, he must have been thinking negative thoughts, and so whatever befell him was his own fault: the victim gets blamed, adding insult to injury. Even the positive side of this supposed law can be pernicious because it easily leads us to ignore the people who in fact deserve to be praised for the good that happens. In Representative Gifford's case, for instance, one would think that the primary tribute should be paid to the able doctors who operated on her and to those who later helped her through her difficult rehabilitation.

Of course the so-called power of positive thinking is not a new idea, and *The Secret* is simply the latest in a long procession of snake oil sales of this particular kind. The early part of the nineteenth century, for instance, brought us the "mind cure," later in that century we had the New Thought Movement, and in 1952 *The Power of Positive Thinking* by Norman Vincent Peale was published around the same time that people thought that women get breast cancer because they are sexually inhibited! I could go on with countless more examples, but the point is that arguably millions of people have been duped over the years into thinking that their minds have magical powers over matter and then swindled not only of their money but in many cases of the power to make rational decisions concerning their lives.

This book is for people who are weary of nonsense and want access to the best that science and philosophy can tell us about our problems. So what does the science of willpower say we actually can and cannot do? One of the most fascinating discoveries in the cognitive science of willpower is that we apparently have a short (if refillable) supply of it and we need to be parsimonious with it, spending it only on things that really matter. For instance, in a simple experiment, subjects were asked to solve a puzzle after having been divided into two groups: the first group was given a chocolate cookie, the other some radish. The subjects in the group that ate the cookies did significantly better than those in the group that ate radish; subjects in the latter group had part of their focus taken away by the exercise in willpower that was required to "enjoy" the radish. Similarly, if you are meeting someone for lunch and are concerned about controlling how much you eat, you may want to go straight to the restaurant without window-shopping at the mall nearby. After using up some of your willpower to refrain from buying yourself a new dress or pair of shoes, it will be particularly difficult to resist putting the butter on that large slice of bread.

The list of ways to deplete our already somewhat meager reserves of willpower is long, and it unfortunately includes many common events in our everyday lives: controlling our appetite, declining to drink (or to drink one more), not having sex (when we want it), suppressing our emotional responses (particularly anger), and even taking a simple test like the puzzle-solving exercise. The first line of defense therefore is to stay mindful and try to avoid having to resist more than one of these (or any other) temptations at any given time. Luckily, cognitive scientists have also discovered something that Aristotle intuited twenty-four centuries ago. (See the discussion of eudaimonia and particularly of akrasia in Chapter 5.) Just as the Greek philosopher suggested that "virtue" is a matter of mindful practice, so modern scientific research tells us that we can improve our willpower. The way to do this is to treat willpower like a muscle (this is obviously only an analogy, not to be taken literally!): you can exercise it by exposing yourself to small temptations and successfully resisting them. For instance, you go out for dinner with friends, and they order dessert. You smile at the waiter, order coffee or tea instead, and then calmly sip your beverage while telling your brain not to salivate too much at the view of the decadent chocolate dessert enjoyed by your table neighbor.

There are also more unusual ways to improve your willpower—for instance, by exercising your body regularly, or by forcing yourself to use your nondominant hand to perform simple tasks. From a physiological perspective, it seems that willpower is influenced by something as basic as the amount of sugar in your blood. Indeed, there is evidence showing that the mere act of exercising willpower reduces blood sugar, which implies that giving a quick boost to your blood sugar level—for instance, by eating a cookie before a test—is likely to give you the edge you need. (Of course, the problem with that strategy is that complex carbohydrates are themselves addictive and not particularly good for your long-term health, and if you become addicted to them you'll then have to rely on your willpower to fight that addiction. Who said life is fair?)

Then again, you could take the religious route. Research conducted by Michael McCullough and Brian Willoughby at the University of Miami clearly shows that so-called intrinsic religiosity (as opposed to the extrinsic variety practiced by people who simply go to church to impress their neighbors) is a good predictor of your ability to engage in self-restraint. This shouldn't really be surprising, considering how many modern religions—particularly those in the monotheistic Judeo-Christian-Islamic tradition—are founded on the idea of exercising willpower to overcome temptation. Of course, it is hard to tell whether engaging in religious practice improves one's willpower, or whether a certain type of person with well-developed willpower is attracted to and can endure religious rituals. Psychological research shows that both adults and children who are religious are capable of more restraint in their actions, and neurobiological data show that the same areas of the brain that are pivotal to self-control are also activated during prayer. Again, however, it is hard to establish the direction of causality.

Interestingly, being spiritual—as opposed to explicitly religious—is not enough: people who consider themselves "spiritual but not religious" (as the by-now-standard phrase goes on so many online dating sites) are no better at exercising their willpower than the rest of us; the implication is that there is something in the religious experience itself, or perhaps in the social setting provided by a religious community, that is efficacious. Still, not all is lost for the secular among us: the researchers from the University of Miami mentioned earlier think that nonreligious people can get the same benefits by engaging in meditation, the secular equivalent of prayer, and by joining secular analogs of churches, such as organizations devoted to social causes. Or you could simply keep up your gym subscription and occasionally write with your other hand.

All of this discussion of willpower is predicated on the basic idea that we actually have a conscious "will" of some sort. Although this notion is, of course, extremely commonsensical, both science and philosophy have a habit of wreaking havoc with our common sense, and sure enough, the very concept of free will is one of those perennial topics of debate, a debate that has been made only more intriguing by the latest contributions from neuroscience. Let's start with the basics on the science side. A major area of the brain involved in our decision-making is the parietal cortex. How do we know this? Because of the sometimes devilish way in which neuroscientists do their experiments. Researchers can stimulate the parietal cortex with low-level electrical currents, and when they do so, subjects report the desire to engage in particular actions—say, rolling their tongue. When the researcher turns up the dial on the electrical stimulation, the subject actually *does* roll her tongue! The experimenter has essentially succeeded in remotely taking over the subject's will, triggering an action that the subject wants to do not because of an internal motivation but because of the electrical stimulus imposed from the outside.

If that's not weird enough for you, consider the ways in which we can now also determine what makes people feel that they "own" a particular action. Was it really I, you might ask, who decided to roll my tongue? The way your brain deals with this is through the parietal cortex's ability to send signals to the premotor cortex, where your movements get started. You feel that the movement was really your own only once your premotor cortex signals back to your parietal cortex that the movement has indeed been executed. Notice that only part of this process deals with the actual action, and that a large part has to do with your desire to engage in that action, as well as your sense of ownership of the action, once performed. Your subjective sense of free will resides in those components, and it can be perturbed if your brain is damaged anywhere in the relevant areas.

But is free will a real ability to make conscious autonomous decisions, or is it rather an artifact of the way the brain works? In other words, are all our decisions made at a subconscious level, before we are even aware of what is going on? This possibility was raised by a now-classic experiment that was first conducted by Benjamin Libet and his colleagues in the 1970s at the University of California at San Francisco and has been repeated and confirmed several times since. It is important to understand exactly what Libet did to see the relevance of these experiments to our discussion. He asked his subjects to carry out the simple action of pressing a button, as many times as they wished within a given time frame. He also asked the subjects to note the exact time when they felt the "urge" to press the button—that is, approximately when they became conscious of wanting to do it.

Libet then measured the interval between the moment in which subjects were conscious of having made a decision to press the button and the moment in which they actually pressed it. On average the time delay was about 200 milliseconds. (Libet had shown that the subjects were reporting the time of their awareness of the decision within a reasonable margin of error of about 50 milliseconds.) So far, nothing extraordinary: subjects made a decision about an action, and it took about 200 milliseconds for their brains to communicate the decision to their muscles so that the action could be carried out. But here is the weird part: Libet and his colleagues also measured—through an electroencephalogram—when the secondary motor cortex registered activity correlated to the decision to act. The secondary motor cortex is the part of the brain that conveys the initial message that eventually leads to the contraction of the muscles. To everyone's surprise, the researchers measured activity in the secondary motor cortex about 300 milliseconds before the subjects said that they had made the conscious decision! Indeed, more recent experiments have shown that there can be a delay of up to seven seconds between the onset of activity in the parietal and prefrontal cortices and the moment in which subjects think that they have made their conscious decision. To put it bluntly, it looks like the conscious so-called decision is an afterthought, a matter of simply becoming aware of the real decision that was already made several seconds earlier by a subconscious part of the brain!

Does Libet's evidence pretty much dispatch the intuitive—and strongly emotionally held—idea that we are the captain of our own (mental) ship? Not quite. First of all, Libet himself did not draw this conclusion from his work. He suggested that while conscious will may not originate the decision to push the button (or do whatever else), it can still "veto" that decision after the motor cortex has given the signal. Of course, this means that our veto power has to be exercised in less than 150 milliseconds, the time it takes for the spinal motor neurons to be activated and to initiate the action. Libet thinks that we all experience this veto power in action, and that this in itself leaves enough room for the concept of a conscious will.

Philosopher Adina Roskies and others go much further than that, however. They point out that Libet's experiments—as interesting and even somewhat disturbing as they are—are concerned with a very narrow and artificially constrained aspect of conscious will. For instance, the subjects in the experiment were not asked to do anything like what we normally associate with deliberation: they were not instructed to explicitly lay out options and provide reasons to pursue one course of action or another. Indeed, one could argue that Libet's experiments do not address conscious will at all, since the subjects were simply told to report when they felt the urge to push the button. It is entirely possible that what Libet measured was just the time it takes for a subconscious urge to come to our awareness. If that is the case, then it is not surprising at all that the activity in the secondary motor cortex can be measured before we consciously know what is going on. This is no different from someone moving rapidly to avoid an obstacle before he is actually aware that there is an obstacle and that he has engaged in an avoidance maneuver. (We will see other examples of this type of "zombie" behavior in the next chapter.)

Although it is certainly the case that philosophers have been talking about free will for a long time without reaching a consensus on how it works, we have to remember that the purpose of philosophy is not to answer empirical questions (we've got science for that, and it does an excellent job at it!), but rather to clarify our thinking. Let's see if a philosophical examination helps us a bit. David Hume famously defined free will as "a power of acting or of not acting, according to the determination of the will"; in other words, free will is the ability to act according to our considered desires, or as Timothy O'Connor of Indiana University puts it, "the ability to select a course of action as a means of fulfilling some desire."

The main problem facing philosophical discussions of free will is the issue of determinism. Simply put, the question is: if everything that happens in the universe is the result of causal necessary relations (for example, the laws of physics), then how can one have a "free" will in the sense of a decision-making mechanism that is independent of influences that are both external (such as environmental ones) and internal (such as one's genetic makeup)? The three types of answers to this question divide philosophers into three camps: compatibilists, libertarian incompatibilists, and deterministic incompatibilists. (Note that the word *libertarian* here has nothing to do with the meaning it has in American politics.)

Compatibilists think that the universe is deterministic, but they think that this in itself does not preclude free will. *Libertarian incompatibilists* think that the universe is not deterministic, but that if it were that fact would preclude free will. And *deterministic incompatibilists* think that determinism is real and therefore free will is precluded. Now, whether the universe is deterministic or not is a question that can be informed by science, since the universe is empirical in nature and current science does seem to have given us a pretty clear answer at this point—though, alas, it is one that doesn't help the debate. If most current interpretations of quantum mechanics (the most accurate of physical theories proposed so far) are correct, then the universe isn't deterministic because there are truly random events (uncaused and completely unpredictable) at the quantum level. Many neurobiologists and some philosophers have seized on this to claim that therefore quantum mechanics provides a scientific answer to the issue of free will. Unfortunately, this is nonsense on stilts, so to speak. Even if quantum events might conceivably "bubble up" to the much more macroscopic level at which the chemical and electrical processes of the brain take place, thus influencing what we do, this would be an example of "random will," not free will. Nobody associates freedom of choice with random decision-making, as if our brains were a roulette machine picking whatever course of action corresponds with a random draw of the wheel. No, we need some other way to think about the debate between compatibilists and incompatibilists.

A famous example of a libertarian incompatibilist is the French existentialist philosopher Jean-Paul Sartre, who said, "No limits to my freedom can be found except freedom itself, or, if you prefer, we are not free to cease being free." As much as I have a certain degree of sympathy for existentialism (its radical doctrine of freedom and consequent responsibility for one's life are intoxicatingly empowering), this simply won't do. There are plenty of limits to our freedom, imposed both by the (physical as well as cultural) environment in which we grow up and in which we live and by our genetic makeup. There is pretty clear evidence from biology and cognitive science that accidents of the chemical machinery of our bodies, for instance, affect our character and our behavior and pose both cognitive and physical limits to what we can do. A clinically depressed person, for instance, can hardly "choose" without constraints not to be depressed and is only partially responsible for whatever decisions and actions characterize his life. Indeed, existentialism runs dangerously close to the same pernicious mentality of blaming the victim that we have seen associated with *The Secret* and other forms of "positive thinking."

The second kind of incompatibilism accepts that the universe is deterministic and denies the possibility of free will. A slightly weaker version of it takes into account the quantum loophole I mentioned earlier, but maintains that—save for truly random events—a science-informed view of the universe, where everything has a physical cause, is simply not reconcilable with any meaningful sense of free will. It follows that, for instance, writing this book wasn't really my choice; it was the foregone outcome of everything else that had happened to me throughout my life up until the point at which I decided to write it. The same, of course, holds for your decision to read it, or for anything we "decide" to do, from the big decisions (choosing a career, getting married) to the very small ones (getting up to get a beer from the refrigerator). The obvious casualty of this type of incompatibilism is any meaningful notion of moral responsibility—or any type of responsibility for that matter. If you really didn't have a choice in doing what you did, then you can neither be blamed (for someone's murder, for instance) nor justly be praised (say, for being faithful to your spouse).

This pessimistic view runs counter to our very strong intuition that we do make decisions that are actually ours in a strong sense of the word. But of course, history is full of instances where common sense flew out the window—just think of the previously universally accepted precept that the earth is flat and at the center of the universe. Still, the trouble for the incompatibilist is that we need only scratch the surface before it becomes increasingly unclear what he means by free will. If "free" here means "uncaused"—that is, completely disconnected from any physical phenomenon or psychological process—then the concept risks running into one of the worst things that can happen in philosophy: incoherence. It is simply hard to make sense of what an incompatibilist might mean when he talks about free will if "free" does not mean "random" and yet also means "unaffected by external and internal causes." Where exactly would such a magical property come from, and how would it work?

This leads us to what the savvy reader has already imagined is my personal favorite: compatibilism. The compatibilist acknowledges that our actions have to be caused, and that they are limited or channeled by physical, biological, and psychological constraints. But the compatibilist claims that this is the kind of free will "worth wanting," in the phrase of Daniel Dennett. No magical hand-waving here, invoking a free will that cannot be, nor the simplistic, existentialist-like rejection of the reality of being human. Quite simply, free will is in this sense our (demonstrable) ability to consider information, balance it against our desires, and take a particular course of action among several available to us. Of course, our desires are themselves the result of our upbringing, our genetic constitution, our experiences in life. How could it be otherwise? And of course, our way of reasoning is also the result of all those things. Again, what would it mean if that were not the case? So compatibilism is a compromise between the undeniable fact that we are a particular type of biological being, with all that entails, and our sense that we own our decisions and can therefore—within limits—be held responsible for them or praised for them.

As in many other areas we are discussing in this book, debates about free will are excellent examples of how both philosophy and science contribute to our understanding. Philosophy helps with clearing up the conceptual issues, and science settles (if possible) the empirical ones. There are some things, however, that science cannot settle, despite some scientists' misguided pronouncements to the contrary. For instance, a recurrent discussion in neurobiological papers concerned with free will issues is that the idea of determinism could in principle be tested by checking whether certain classes of brain signals follow a pattern that is clearly random. This proposal is conceptually confused on several levels. First off, any set of empirical data may look random until we discover the causal mechanism generating it—that is, "randomness" is often simply how we label our ignorance of a given phenomenon. Second, as I explained earlier, even if it were possible, showing beyond reasonable doubt that some brain events are truly random would not purchase us free will in any meaningful sense of the word. So, while neurobiology certainly has a lot to tell us about this issue, one thing it will not do is settle the debate between compatibilists and incompatibilists.

Another common mistake in discussions of neurobiology and free will is to conclude that if science could give us a mechanistic account of the phenomenon, then in some important sense science would have demonstrated that free will doesn't exist and that it's all about neural pathways. Again, this claim strikes a philosopher as bizarre. *Of course* free will, however conceived, has to have a neural basis of some sort. Unless we are talking about magic, *everything* that human beings feel and think will turn out to be based in their brains in one way or another. No brain, no feelings or thoughts. So providing a mechanism for phenomenon X does not in any way tell us that X was some sort of illusion; it simply tells us how it is that X can be part of the human experience, because human beings are biological organisms that require a physical substrate to have any experience.

It also turns out that there is more than one conception of free will that can be investigated neurobiologically. Specifically, neurobiologists distinguish among at least the following five possibilities: (1) free will as the initiation of motor activity, as in Libet's experiment; (2) free will as "executive control," that is, Libet's idea that we still have veto power over our unconscious decisions; (3) free will as a feeling of ownership, which we have seen has its own neurological basis; (4) free will as intention, which philosophers think of as a representational stage between deliberation and action (though, according to some, intentions may be unconscious); and finally, (5) free will as decision-making, which can be a long process that takes hours or days, depending on its object. That neurobiologists have identified at least these five aspects of free will as fodder for research suggests the very real possibility that what we think of as free will isn't a unitary phenomenon after all, but a broad label we apply to a set of disparate things that the brain does. And so the interplay between philosophical clarification and scientific investigation continues.

CHAPTER 10

WHO'S IN CHARGE ANYWAY? THE ZOMBIE INSIDE YOU

> Reason is, and ought only to be the slave of passions, and can never pretend to any other office than to serve and obey them.
>
> —DAVID HUME

YOU ARE ABOUT TO LEAVE YOUR HOUSE, BUT BEFORE you open the door you consider whether you should bring an umbrella with you or not. There is a perfectly rational way to assess the likelihood of rain based on the best available information. For instance, let's assume that you've noticed dark clouds outside and that a bit of research (probably on the Internet) shows you that whenever it rains there is a 90 percent chance that that type of cloud is around. You also find out, however, that when it does not rain, there is still a 30 percent chance of seeing the same type of clouds in the sky. (Notice that these are not complementary events, so their probabilities don't add up to 100 percent.) Should you take the umbrella?

Your intuitive answer is probably yes, and you would be right. Statistical theory shows that the proper, formal way to assess the chance of rain in this situation is to take the logarithm of the ratio of the two probabilities: $\log(90/30) = 0.48$. Since this number is greater than zero, you should in fact pick up your umbrella on your way out. Had the probabilities been the other way around ($\log(30/90) = -0.48$), the result would have been a negative number, so the odds would have been against rain and your best bet would have been to leave the umbrella at home. (Unless you have a phobia about getting wet, in which case you should carry an umbrella with you at all times.)

But surely nobody in his right mind would engage in these or more complex calculations before making such a simple decision as whether to pick up an umbrella? You would be surprised. This example was put together by Paul Cisek of the Department of Physiology at the University of Montreal to explain an interesting finding of recent neurobiological research: it turns out that your brain has neurons that specialize in precisely this sort of calculation, except that you are not consciously aware of what is going on. It is as if a zombie inside you is in control of your operations and decisions, and "you" (meaning your conscious self) realize what is going on only after the fact.

The research that Cisek was commenting on, published by T. Yang and M. N. Shadlen in *Nature* in 2007, was conducted on monkeys instead of humans, and it had to do with how the animals interpreted symbolic information—information exactly analogous to your knowledge of the probability of rain when you observe certain clouds. The monkeys were exposed to two targets, a green one and a red one, and one of them was associated with a reward. The animals had to guess which target to choose based on cues provided in the form of geometrical figures that were probabilistic predictors of a given reward. For example, a triangle appeared in only 5 percent of instances where the red target was correct (that is, when it was associated with the reward) but in 50 percent of instances when the green target provided the reward. If the monkeys were capable of what philosophers call "probabilistic inference," they should have regarded triangles as a cue indicating green targets. Not only was that in fact the case, but the researchers showed that when more complex cues were given, the monkeys behaved as if they were deploying the concept of the logarithm of the likelihood ratio—the very same one you used to decide whether to pick up an umbrella on your way out the door.

But monkeys don't know about logarithms. Indeed, many human beings don't know about or understand logarithms—let alone the theory behind probabilistic inference—so how is this possible? It's possible because your brain does it all for you (and the monkey) without any need of your conscious awareness! Yang and Shadlen showed that there is a very tight correlation between the activity of certain neurons (called LIP, for lateral intraparietal area, the area of the brain where they are located) and the logarithm of the likelihood ratio of the cues provided during the experiment. That is, the higher the value of the logarithm, the stronger the activity of the LIP neurons, as if the monkeys had a built-in inferential calculator in their brains that allowed them to use the available information most efficiently and to get the reward more times than not.

To some extent, all of us become aware of having a "zombie moment" at one time or another. A baseball player hitting a fastball pitched at ninety miles per hour does not have time to solve a complex set of differential equations to tell him where and when to swing the bat. And yet his brain seems to do the calculations for him without conscious input. Indeed, if consciousness were required in baseball, the game would be even slower than it already is, because it takes hundreds of milliseconds (an eternity in most sports) for consciousness to get the body into swinging mode.

I used to live near the Brooklyn Bridge, on the Brooklyn side, and often had to cross a double pedestrian light before reaching my apartment. The light farthest from me always turned green for pedestrians before the near light (because of the convoluted traffic patterns at the entrance of the bridge). Although I had to wait for the second light to turn green before I could safely cross, I often caught myself with a foot already off the sidewalk, as if an internal autopilot had seen the first light going green and had automatically (though erroneously, in that case) equated green with "go." Usually I became conscious of this mistaken (and potentially very dangerous) decision in time, stopped my leg from moving further, and patiently waited for the second light to turn green before crossing.

This experience illustrates an idea of Libet's that he developed in the years after his famous 1983 experiment (see Chapter 9): perhaps the role of consciousness is not to engage in constant evaluation and decision-making, because it would be too inefficient to do so in most practical situations, but rather to monitor the internal zombie's activities (it becomes "aware" of its output) and occasionally exercise a veto or redirect actions to improve over the zombie's fast, but sometimes inaccurate, decisions.

Our autopilot zombie, then, can be very useful, especially when it comes to complex operations like hitting a baseball or calculating the odds of rain. The problem is that the more we learn about our subconscious decision-making mechanisms, the more we find out that they can be easily manipulated by others, without our notice. We encountered an example of this phenomenon, called "priming," in Chapter 6 when I told you about the experiment showing that thinking of the last digits of your social security number (if they're high) makes you more prone to overpay for a given item. The same phenomenon was at play when another set of researchers showed that the outcome of a job interview (or a date) is affected by holding a cold or a hot liquid (engendering "cold" or "warm" reactions toward the interviewee or the date).

There is direct neurobiological evidence that a lot of our decision-making is done by subconscious processing of information, even when we *think* we are processing information consciously. In an experiment published in *Science* magazine in May 2007, a group of neuroscientists scanned the brains of players of a video game. The participants had to squeeze a handgrip whenever they saw the image of some currency on the screen, and they were told to squeeze it tighter the more valuable the currency was (for example, a pound note versus a penny). The interesting twist to the experiment was that some of the images were visible long enough to register consciously, while others flashed rapidly in and out of the screen and were perceived only subliminally. In both cases, however, the same region of the brain, the ventral pallidum, was activated when the subjects squeezed the handgrip. This is surprising because the ventral pallidum is an area of the brain that is evolutionarily very ancient and is not involved in conscious thinking. The implication is that the reactions to both subliminal and conscious imagery were decided subconsciously. The prefrontal cortex, where conscious thought originates, was literally the last to know.

The idea that there are several components to the human mind, operating in a quasi-independent manner that sometimes brings them into conflict, is of course not new at all. In 1920 Freud published an essay entitled "Beyond the Pleasure Principle" in which he presented his theory of a tripartite mind, with the parts labeled "id," "ego," and "superego." The id for Freud was the seat of sensual drives, particularly but not exclusively sexual ones. The superego represented the conscience and embodied social norms. The ego was the intermediary between the other two, mediating the constraints imposed by external reality. Freud's ideas had no empirical neurobiological basis (and today's research shows a much more complex picture of the relationships between conscious and unconscious minding) and were instead rooted, surprisingly, in philosophy.

Perhaps the most ancient theory of the mind as made up of multiple interacting parts is Plato's concept of the "soul," as presented in two of his dialogues, the *Phaedo* and the *Republic*. The two dialogues present two versions of the theory, so here I will briefly discuss the one articulated in the *Republic*, since it is the more mature version and the one that is pertinent to our discussion. Plato's aim in the *Republic*, as the title may suggest, was not really to investigate how the human mind works, but rather to pursue the question of how we should build an ideal state. Yet, the Greek philosopher draws a direct parallel between the state and the individual human beings who are parts of the state, suggesting that a just state is made possible by the harmonious balance of its component parts, just as a happy person is the result of a balance achieved among the components of his soul. Hence the somewhat strange idea of analyzing what makes for a balanced soul in order to draw conclusions about what makes for a just state.

The concept of "soul" in ancient Greece was varied and complex and did not necessarily include the idea of survival after death. (It did for Plato, but not for his pupil Aristotle.)

For the purposes of our discussion, we can roughly equate the soul that Plato was talking about with our (sometimes equally fuzzy) idea of mind. For Plato, the soul/mind has an appetitive part, a spirited part, and a rational part. The appetitive component is concerned with the satisfaction of fundamental instinctual desires for food, water, or sex. In this sense, the appetitive soul is directly analogous to Freud's id. The spirited part of the soul deals with self-preservation and is also the source of courage (on the positive side) and anger or envy (on the negative side). The rational component, as the name implies, is the seat of wisdom and higher-level thought and is concerned with issues of truth. The analogy between the spirited and rational parts of the soul, on the one hand, and Freud's ego and superego, on the other, is a bit less obvious, though the parallel is not that far-fetched either.

The reason it is interesting to consider Plato's view of the soul, however, lies not in the details of the philosopher's tripartition, but in the idea that the various components are often in conflict with one another and that there is a hierarchy of importance among them. For Plato, a good life is possible only if the three parts of the soul are in equilibrium, and this equilibrium is most certainly not a democracy (and not surprisingly, neither is the ideal state that he ends up advocating in the *Republic*). Instead, for Plato the rational soul ought to be in charge and keep the appetitive one in check, helped in this by the spirited soul. Plato compares the soul to a chariot: a wild black horse represents the appetitive soul, flanked by a more noble white horse (the spirited soul), and both are under the stern command of the charioteer (the rational soul). So here we have a philosophical—and colorfully presented—theory of what makes for a balanced human being: such a human being is one who cultivates reason over passion as a guide to her life. In modern terms, if we can stretch our interpretation of Plato a bit, the philosopher would claim that conscious thinking ought to guide and keep in check subconscious instincts. As we have seen, however, although the higher brain functions have some room to exercise their veto power over the zombie inside us, it is beginning to look like it is Plato's horses that steer the charioteer, not the other way around.

But perhaps this state of affairs is not such a bad thing. Another influential philosopher, David Hume (who wrote some twenty centuries after Plato), turned the cards around and claimed not only that "reason alone can never be a motive to any action of the will" but most famously that "reason is, and ought only to be the slave of passions, and can never pretend to any other office than to serve and obey them." Hume was no defender of irrationality. Indeed, he was friends with the French philosophers of the Enlightenment (also called the Age of Reason), and he is remembered for his exquisitely well-argued writings on matters of morals, politics, and science. What, then, could Hume possibly mean by saying that reason not only *is* (as a matter of fact) but *ought to be* the slave of passions? Hume was a keen observer of human nature, and he realized that we do things because we have motivations, but that motivations come out of "passions" (emotional drives), not reason.

Suppose I stop typing this on my laptop's keyboard, get up from my chair, go to the refrigerator, open the door, pick up a bottle of water, and start drinking. I don't do all of this because my reason is telling me that if I do not I will eventually get dehydrated, lose my concentration, and possibly die of thirst. I do it because I'm *thirsty*—that is, I have a feeling of thirst that was registered by my brain and the decision was made by my inner zombie to act on it. In this scenario, there's no need to invoke conscious, rational decision-making.

Hume generalized this principle to apply not just to obvious instances, such as my little example here, but to pretty much anything of consequence we do as human beings. As he put it: "'Tis not contrary to reason to prefer the destruction of the whole world to the scratching of my finger. 'Tis not contrary to reason for me to chuse my total ruin, to prevent the least uneasiness of a ... person wholly unknown to me. 'Tis as little contrary to reason to prefer even my own acknowledg'd lesser good to my greater, and have a more ardent affection for the former than for the latter." As counterintuitive as Hume's ideas are, and as much as they went in direct opposition to a long tradition in philosophy stemming from Plato and continuing to our day, modern neurobiology seems to vindicate Hume when it portrays reason as the instrumental tool deployed to achieve our desires, with the fundamental engine generating those desires lying much lower than the cerebral cortex.

Hume went a step further and famously maintained that morality itself is not arrived at by logical argument (again, contrary to what most philosophers had argued before and have since), but is rather the outcome of our emotional reactions. Let us consider an example. Most people today feel that slavery is a repellent and morally wrong practice, but of course that has not always been the case in human history. Now, one can in fact build a logical argument against the practice of slavery to show that it is wrong from a rational perspective. But such an argument would have to rely on certain premises that are not, in themselves, easy to defend on rational grounds. For instance, one could say that it is wrong to limit the freedom of other human beings, or that we should not force on others what we would not want others to force upon us. Yet a hypothetical defender of slavery could counter such arguments with a logic of his own: that it is rational to limit some people's freedoms so that a stronger and more prosperous society can be built, or that it is acceptable to force others to do our bidding if we have the might to impose it on them, and so on. You may not find such proslavery arguments convincing (I certainly hope you don't!), but the point is that one can rationally argue both sides of the debate and that ultimately our moral sense derives from how we feel about slavery, with our arguments elaborating on that feeling, not determining it.

Earlier (in Chapter 4), we encountered the idea that morality emerged out of an evolutionary process that first shaped our ancestors' instincts (and hence their emotional reactions) and only much later on developed into the complex practices of modern human societies. Hume obviously was not thinking of evolution (he wrote before Darwin), but he would have probably been very pleased with the idea nonetheless. In fact, another intriguing example of modern research in cognitive science shedding some light on philosophical concepts manages, once again, to vindicate Hume. A study published in *Science* magazine in February 2009 and conducted by H. A. Chapman, D. A. Kim, J. M. Susskind, and A. K. Anderson made a surprising connection between physical disgust and moral disapproval, hinting that there might be a deep connection between basic human emotions and our more sophisticated moral judgment. The authors showed that a primary muscle involved in expressions of disgust elicited by bad tastes or smells or by disease is also activated when we experience moral disgust—for example, when we are treated unfairly.

This so-called oral-to-moral hypothesis is still a bit speculative, but it is in agreement with a key prediction of evolutionary theory: evolution builds on previously existing mechanisms and structures and recycles them for new functions. In this instance, the primordial response, common to many mammals, is a reaction of disgust to the experience of bitter foods, which in nature are often poisonous. The idea is that the evolution of moral reactions was facilitated by co-opting this preexisting "oral" route to disgust and using much of the same brain and muscle apparatus to express rejection of complex biologically dangerous (and hence morally reprehensible) behaviors like incest. The last step would then have been to co-opt again the same physiological machinery for the expression of morally even "higher" levels of disgust, such as those elicited by discriminating and unjust behavior (say, being shortchanged in a financial transaction).

The connection between deliberative reasoning and emotional impulse, which Plato thought ought to be supervised by our rational self and Hume concluded was instead controlled by our passions, can go wrong when people pathologically succumb to their impulses. About 9 percent of Americans, it turns out, have a problem with compulsive behavior that leads them to make rush decisions about their lives that they are likely to regret, to have trouble planning for their future, or to engage in self-destructive behaviors like drug and alcohol abuse.

Neurobiologists know that impulse suppression is located in an area of the prefrontal cortex of the brain known as the dorsal anterior cingulate. You can think of it as the brain's braking system, and it does not mature completely until the end of adolescence—which can go a long way toward explaining what cognitive scientists call "risk-taking" behavior in young adults. Interestingly, there are genetic effects on the ability of the dorsal anterior cingulate to do its job. For instance, remember MAO-A, the gene associated with psychopathology that we encountered when we examined the story of Jim Fallon in Chapter 3? It produces an enzyme that reduces the brain activity of serotonin, a neurotransmitter that affects our mood (including how hungry we are, or how angry). A variant form of MAO-A has been linked to excessive impulsive behavior, leading neuroscientists to perform brain scans of subjects with the normal version of the gene and of others with the high-risk version while they were engaged in a video game that tested their propensity for impulsive decisions. Intriguingly, people with the high-risk variant of MAO-A also showed lower activity in the very same dorsal anterior cingulate that keeps impulsive behavior in check.

Then again, the problem is not all in our genes—far from it. The same effect of letting go of the brain's brakes, thereby succumbing to emotional impulses, can also be brought about by external or environmental circumstances that we can exercise only partial control over, if any, such as stress or alcohol and drug abuse. In a further twist that is scientifically intriguing but makes helping people much more difficult, external conditions interplay in multiple ways with genetic ones to generate what scientists call "gene-environment interactions."

All of this complexity notwithstanding, we should by now have gained a dose of humility about how much in charge of our lives we really are, at least if by "we" we mean our conscious self. It will come as no surprise in the next chapter, then, that things are particularly messy in the conscious-versus-unconscious department when it comes to one of the most vital and more meaningful components of our emotional lives: love.

PART IV

LOVE AND FRIENDSHIP

CHAPTER 11

THE HORMONES OF LOVE

You can't blame gravity for falling in love.
—ALBERT EINSTEIN

THE GOD OF LOVE LIVES IN A STATE OF NEED, WROTE Plato in the *Symposium*, a dialogue on the various forms of love in which we get sex advice from none other than Socrates! Philosophers, and of course novelists and poets, have been writing about love for millennia for the simple reason that it is a fundamental emotion that profoundly affects our lives. But precisely because love is a universal phenomenon in our species, one may wonder about its evolutionary origins, and because it is an emotion, we can also ask what mechanisms in our brain make it possible. And those questions lead us into the still young and already controversial science of love.

How can anybody seriously entertain the thought of putting such a complex human emotion as love under the scientist's microscope without ending up looking ridiculous, or at any rate missing the essence of what is going on? For starters, we can use smelly T-shirts. In 1995 Claus Wedekind and his collaborators at the University of Bern in Switzerland published a now-famous paper in the prestigious *Proceedings of the Royal Society* in which they claimed that human females display clear preferences for males who smell a certain way. The researchers followed a simple protocol: they told a number of men to wear the same T-shirt for a couple of nights; then they asked a group of women to smell the shirts and rate the odors they perceived for sexual attractiveness. This sounds like a bunch of crazy scientists run amok, but there was a logic to the experiment: Wedekind and his colleagues knew that other mammals (for instance, mice) express olfactory preferences for potential mates, and that the reason for this is that an individual's smell is related to the genes he carries at the major histocompatibility complex (MHC)—an important molecular tool of the immune response system through which our bodies defend themselves from external attack by pathogens.

So the Swiss scientists scored both the men and women in their experiment for their MHC molecular markers and then compared the data with the women's preferences based on the smell of the T-shirts. The results were astounding: humans behave like mice, with the females demonstrating a significantly stronger preference for males who sport MHC genes different from their own. This makes perfect sense from an evolutionary perspective, because when people with different MHC genes mate, their offspring will have more genetic variants at their own major histocompatibility complex, which in turn increases the chances of that offspring surviving an infection. (It's like having a wider array of defense weapons at your disposal: if the enemy disables one kind, you can always deploy another one.)

This is an intriguing example of how an evolutionary prediction (parents should try to maximize genetic variation in the immune response of their offspring), already borne out in the case of animal systems (mice), turns out to predict the behavior of otherwise much more complex organisms such as ourselves. Next time you are on a date, it might be good to get close enough to your potential partner to smell him and see what sort of reaction you get. (Assuming, of course, that your main goal is to have healthy children; should your pursuits have more esoteric goals, like finding a companion who can make you happy, things become much more complicated.) One piece of cautionary advice, however: Wedekind and his colleagues found that women's ability to discriminate among men with different types of MHC disappeared if they were on a contraceptive pill. Apparently the pill's alteration of the woman's hormone balance in some way interfered with her ability to pick up on subtle smells, making it impossible to express a preference related to the MHC. So the best thing to do is to go on a date without chemical distractions (not just the pill, but perfumes as well), and possibly after not having showered for a couple of days.

Of course, love cannot be reduced to a simple matter of smells and hormones (despite the fact that the MHC study does give a whole new meaning to the idea of "chemical attraction" between people), and philosophers have been discussing the idea of love and its implications for human affairs at least since Plato. The ancient Greeks distinguished at least three fundamental types of love. *Eros*, of course, is what we today call erotic love, which has largely to do with sexual attraction. For Socrates, eros is "incomplete" because it is characterized by constant dissatisfaction, a search for the other that can never be but temporarily fulfilled (though more on why this is and how our brain actually manages it in a few moments). *Philia* is the sort of love we experience when we need or want to get along with other people; for Aristotle, philia includes parents, children, and lifelong friends, but also business contacts and political alliances. The last type of love, *agape*, is in some sense the purest: agape is the sort of love that we feel unconditionally and that leads to self-sacrifice—for instance, sacrifice to the gods (if one believes in them), but also sacrifice for a spouse or close family member (in this sense overlapping with the idea of philia), or even for an idea or pursuit, such as love of science or of truth.

Modern philosophers continue to earnestly discuss what love is, and they have proposed four different, though perhaps partially overlapping, conceptions of love that are significantly distinct from those of the ancients: (1) love as an emotion, (2) love as a "robust concern," (3) love as a union, and (4) love as valuing the other. Let us start with the idea of robust concern. The defining feature of this kind of love is selfless interest in the other's well-being, for his or her sake and not because we gain anything out of it. (As you may have noticed, love as robust concern is reminiscent of the ancient Greeks' idea of agape.) In the somewhat dry and formal words of philosopher Gabriele Taylor:

> If x loves y then x wants to benefit and be with y etc., and he has these wants (or at least some of them) because he believes y has some determinate characteristics Ψ in virtue of which he thinks it worthwhile to benefit and be with y. He regards satisfaction of these wants as an end and not as a means towards some other end.

All right, I promise never to quote a technical paper on the philosophy of love directly again, because this is the sort of thing that gives philosophers a bad reputation. Still, what

Taylor is saying is that we don't love the other (y) because her characteristics (Ψ) benefit us, but because they are worth cherishing in their own right. Again, in this sense we are not far from agape, as this is the same sort of love that is supposed to be reserved to the gods because they are inherently good (and that, for instance, you would rightly deny to evil entities like demons). Although this idea of selfless love has some commonsense appeal, there also seems to be something clearly amiss. As another philosopher, Harry Frankfurt, put it (not a direct quote!), the idea is that robust love is neither a matter of feelings nor a matter of opinions, but a matter of will: we love someone in a robust fashion because she acts in accordance with a set of motives and preferences that we approve of. (Think about loving your god because he is good and acts accordingly, and that you would be bound not to love him if he started behaving in an evil manner—how you know whether your god behaves one way or another is, of course, your problem, upon which I shall say nothing further.)

A second modern philosophical view to entertain is that of love as valuing the other person. The basic idea is that love *means* to value someone in himself or herself, and that we do so because of an appraisal that centers on the dignity of that person. If this sounds a bit abstract and detached from the real world, well, it is. But there is an important kernel that philosophers who support the value conception of love are trying to get at: the idea that a love object (a person) cannot simply be swapped for another one with similar characteristics, because this would violate the dignity of both people. Think of the "robust concern" view just discussed: there is nothing in that view that would preclude you from having the same "concern" for (that is, loving) another object with the same characteristics as the one you are loving now. You could therefore swap gods or lovers, or even love many gods and many people at the same time, as long as they share the same set of characteristics (Ψ). Some people might be okay with this, but others feel that real love ought to be more exclusive and less subject to commodification. If you are in the latter group, then the value view of love might fit you well.

The third modern philosophical perspective is of love as a union. This is the idea that what is central to love is two independent individuals forming a third, collective union, a "we" that becomes more important than and transcends each individual "I." Some philosophers speak of this "we" entity in a clearly metaphorical way, while others seem to give a more serious ontological (pertinent to existence) status to the ensemble, almost as if it really were a new individual in its own right. (It is not clear to me in what sense the "we" of love can be anything but metaphorical, but there it is.) As with the value view of love, the union conception tries to capture something that most people who have been or are in love can relate to: the creation of a new set of priorities as the couple as a unit becomes more important than the individuals who constitute it. But therein lies a problem as well: human beings are *both* social and fairly individualistic animals, and one can object that a union view of love puts too much emphasis on the couple at the expense of personal space, rights, and dignity. As we all know, it is precisely this tension between joint and individual needs that often is at the root of relationship problems in real life.

Finally, we turn to the emotional view of love. In philosophy an emotion is a combination of an evaluation of the object of the emotion and a motivational response to that object. For instance, if I'm afraid of you, that means I have evaluated you as somehow dangerous to my health, and it probably also means that I am prepared to take some action against you, either defensive or evasive. Of course, thinking of love as an emotion would hardly be surprising for the nonphilosopher, but the question for us here is: What sort of understanding of the phenomenon can be gained this way? And what potential problems arise if we conceptualize love primarily as an emotion?

One thing that philosophers get out of emotional theories of love is being allowed to distinguish loving someone from simply liking someone. If love is a distinct and deeper sort of emotion than the emotions elicited by friendship or admiration, then we begin to see why those other experiences are so clearly not like love. According to several philosophers who support an emotional view of love, what accounts for much of this difference is that we share a unique narrative history with the beloved: regardless of how he or she will change throughout life, we keep accumulating common memories of events and situations that are obviously unrepeatable with anyone else. This, according to such philosophers, also explains why we don't commonly "trade up" at the first opportunity, why we do not switch partners as soon as we meet someone with even better characteristics (call them "$\Psi+$") than the one we are currently engaged with.

The problem, of course, is that as a matter of regular sociological observation, people *do* trade up (or trade partners at the same level, or sometimes even trade down). And it is certainly the case that shared histories do not stop people from leaving their lovers because of a variety of both external and internal circumstances. People change over time in ways that may not be predictable when a relationship gets started, and the change may amount to a sufficient reason for the two partners to decide that the continuation of the relationship is no longer warranted. Besides, if we are thinking of shared history as a powerful glue holding people together, we need to remember that we also share narratives with other people, such as friends and colleagues, but that this does not prevent us from making new friends or changing jobs.

It seems clear that all four of the major theories of love being discussed by philosophers today have problems of one sort or another. And yet they all also pick up on something right about human relationships; perhaps a clever combination of these views is warranted to begin to achieve a philosophically satisfactory account of love. However, from a sci-phi perspective, no such account could be truly adequate unless it took into account what science has to say on the subject. And if there is anything that scientists are discovering about love these days, it's how multifaceted and complex a biological emotion it really is.

One way to summarize the science of love would be to say that it's (almost) all about hormones and their effect on our brains, and how this in turns translates into the behaviors we exhibit—the whole thing sprinkled with a bit of reasonable speculation about the evolutionary origins of such hormone-brain links. Helen Fisher, an anthropologist at Rutgers University, has made a career of researching and writing about love (and she has turned her eye to for-profit applications of her work by teaming up with a dating website appropriately named chemistry.com). Fisher and her colleagues think that there are three phases of love; that they are characterized by specific neural correlates (different parts of the brain are differentially activated during each phase); that they are regulated by distinct hormones acting in various combinations; and that we share this basic set of processes with many primates and other mammals—or at least that minority of mammals (about 3 percent of species) that form stable couples in order to raise their offspring.

The three phases themselves will be familiar to anyone who has fallen in love in his or her life: we start with infatuation (driven by a sexual interest), continue to romantic love, and—if things keep going along that trajectory—settle into long-term attachment. (Sometimes, of course, the cycle starts all over with another partner after some years.) What is interesting is that Fisher and others have been able to show that each phase is not just defined by externally observable, socially structured behavior but accompanied by specific changes in hormonal activity that act on particular areas of the human brain. Here is your handy-dandy guide to hormonal love for future reference:

PHASE	HORMONAL PROFILE
Infatuation	High levels of androgens, particularly testosterone
Romantic love	High dopamine, low serotonin
Attachment	High oxytocin and vasopressin

Beginning with the process of infatuation, we have to remember that although testosterone is usually associated with male prowess, it is present in both men and women and elicits similar effects in them, at least in terms of sexual drive. This phase is straightforward to explain in evolutionary terms: as Richard Dawkins once aptly put it, every single one of your ancestors, unsettling as the thought may be, had sex at least once or you wouldn't be here. Sex drive, of course, tends to be rather broadly directed, meaning that there are many individuals whom we find sexually attractive and with whom, given acceptable social circumstances, we would have sex.

Pretty quickly, however (how quickly depends on both individual characteristics and social milieu), if the conditions are right, infatuation develops into romantic love, which was certainly not invented by the Victorians. Fisher's own book, *Why We Love: The Nature and Chemistry of Romantic Love*, is peppered with literary citations aimed at showing that the phenomenon is cross-cultural and has been documented since the beginning of human written history. (My favorite is a phrase apparently typical of the natives of rural Nepal: *"Naso pasyo, maya basyo,"* which translates as: "The penis entered and love arrived.") Those of us who have experienced romantic love will tell you that it is a strange phase in the human condition: one becomes literally obsessed with the object of his love, losing sleep over her, always wanting to be with her and only with her. This response, chemically speaking, isn't surprising: dopamine, one of the two chief hormones involved, is the foundation of the so-called reward system of the brain—the very same one that gives us a chemical pat on the back when we do something satisfactory and that also uses the kind of brain receptors that are sensitive to addictive drugs like cocaine. Romantic love literally is an addiction! Moreover, serotonin, the second important hormone during this phase, stays at particularly low levels in the course of romance. Low levels of serotonin are well known to be associated with obsessive behavior, and also with the tendency to behave on impulse. Sound familiar? Well, now you know where it comes from.

Some of the most intriguing results of research on romantic love, however, come from animal systems. Just because they cannot recite Shakespeare's sonnets doesn't mean that prairie voles do not show the exact same obsessive-compulsive behavior toward their mate that human beings do—which is not surprising since the evolutionary aim is probably the same: to convince a partner you are sexually attracted to (see infatuation above) to share her favors with you so that your genes can happily combine and be transmitted to the next generation. Now, suppose you inject a female prairie vole with a dopamine antagonist, a chemical that selectively blocks the brain's uptake of dopamine: the result will be her sudden loss of interest in whatever male she was involved with at the time! The experiment can also be done in reverse: injecting the female prairie vole with a dopamine *agonist*, a substance that facilitates dopamine uptake, will make her interested in the courtship behavior of any male who happens to be nearby when the injection takes effect. Seems like the idea of a love potion (and its antidote) is no longer just a matter for fairy tales (or science fiction horror stories).

The last phase is the calmer type of feeling that replaces the initial high-testosterone sexual drive and the subsequent dopamine-directed romantic obsession: we slip into a comfortable sense of stability and form long-term attachments. Again, evolutionarily this does not take a rocket scientist to explain: stable pair-bonding is typical of animal species for which there is a premium in both parents helping to raise the offspring. This is nowhere more true than in the case of humans, for whom the new generation takes several years to become self-reliant enough for the parents to be able to relax their involvement (and, in early human societies, begin to attend to the task of making still newer progeny). As I mentioned earlier, researchers have found that two hormones play the key role here: oxytocin and vasopressin. Again not surprisingly, they are both well known to be involved in nesting behavior in other species. What's even more convincing is that the brain receptors for these hormones are absent or much less numerous in species, such as rhesus monkeys and white-footed mice, that are promiscuous and do not engage in long-term pair-bonding.

The standard, and somewhat reasonable, objection that most people raise to this sort of chemical-neural analysis of love and other human emotions is that it is far too simplistic: surely the love that inspired Shakespeare's sonnets (not to mention the sage pronouncements of the natives of rural Nepal) is not *just a* matter of chemicals and neural firings. This is of course true, but it misses the point, which is that chemicals and neural firings have a lot more to do with many facets of human behavior than most people realize or are willing to acknowledge. For instance, the sort of research discussed here also sheds some light on what happens when things don't exactly go as they are supposed to, at least in fairy tales (you know, " . . . and they lived happily ever after"). Take the obvious question of why romantic love usually doesn't last. Studies show that typically feelings of romance persist for about twelve to eighteen months, although of course in individual cases that period may be much shorter or longer. (This is neural psychology, not subnuclear physics, so things are true only in a broad statistical sense.) Fisher suggests that romantic feelings don't last because we simply can't keep up with the stress: romantic behavior is what biologists call "metabolically expensive," meaning that it requires a lot of effort to seduce a female (human or not). The deed is accomplished differently by different species, of course, with bowerbirds building elaborately decorated nests (which are then discarded in favor of more functional versions once the female has consented) and human beings buying dinners, diamond rings, and houses, but the concept is the same. (The male bowerbird, incidentally, is simply extraordinary: he goes so far as collecting shiny objects and even paint to "decorate" what has been described as an elaborate bachelor pad, the sole aim of which is to beguile the female.)

Another crucial and decidedly un-fairy-tale-like aspect of human love relationships that we can make sense of by thinking about biology is the striking fact that many such relationships are not monogamous or do not last a lifetime. The statistics are pretty clear: polygyny is present in over 80 percent of human cultures. (Although historically, of course, very few males have been able to afford multiple mates at the same time; see "metabolic expenses" above.) Moreover, surveys show that a whopping 30 to 50 percent of Americans (both men and women, despite cultural myths to the contrary) engage in extramarital affairs at some point or another during their lives. Finally, not only do open societies (where people are more free to follow their choices and inclinations) tend to have high divorce rates, but these rates peak around the fourth year of marriage. Why? A biologist would point out that four years is about as long as it takes for a human child to become self-sufficient enough not to require biparental care anymore. Further, plenty of mathematical models (and empirical evidence in other species) show that there is a premium for both males and females to seek a variety of sexual partners during their lifetime, because doing so increases the chances that their offspring will win the genetic lottery. The natural state for primates like us is one of serial monogamy or limited polygyny. Of course, it does not follow that what is natural is what we ought to do. (As we saw in Chapter 1, that is a well-known logical fallacy in philosophy.) If we choose to go against strong biological instincts, however, we need to be aware of the perils and difficulties of that course of action: it will cost us quite a bit of willpower, and we will be constantly putting ourselves in the position of feeling guilty just because we are thinking (although not acting) along lines that our genes, hormones, and brains laid down as the path of least resistance millions of years ago.

A better appreciation of the chemistry of love also has more immediate, practical consequences. For instance, consider the increasing number of people who take antidepressant medications. A study by the Centers for Disease Control and Prevention (CDC) released in 2007 showed that the use of antidepressant drugs tripled in the United States between 1988 and 2000. In 2005 alone antidepressants were prescribed 118 million times, and in 2004 the total revenues for this type of drug hit the astronomical amount of $14 billion. (Comparable statistics for more recent years were not available at the time of this writing.) Many people, of course, take these drugs for good reasons, and doctors are aware of an array of side effects typical of each drug, effects that need to be weighed against the benefits in each patient's case. But Fisher and her coauthor J. Anderson Thomson Jr., in an article published in 2006, warn of additional side effects that may be below the radar of both doctors and patients. Many antidepressant medications are so-called selective serotonin reuptake inhibitors (SSRIs), which work by blocking the brain from eliminating serotonin, leaving the chemical in circulation for longer periods of time. This is necessary because serotonin has the effect of elevating one's mood, thus countering feelings of depression. But as we have just seen, feelings of romantic love rely on low levels of serotonin; thus, patients on antidepressant medications may experience artificial changes in the way they feel about their companions, regardless of any objective change in their external circumstances.

This is more than a mere possibility, as shown by an increasing number of case studies. A typical one, quoted by Fisher and Thomson, concerns a man who started using medication to counter bouts of depression: "As appreciative as I was to have regained my health, I found that my usual enthusiasm for life was replaced with blandness. My romantic feelings toward my wife declined drastically." Eventually he gradually discontinued his medication (under medical supervision), and his feelings for his wife returned as if by magic. Of course, it wasn't magic, but a rather powerful example of how the help that science gives us in coping with our problems often comes with trade-offs. It is precisely these trade-offs that biologists cannot help us with in making our decisions—which is why we still need philosophers.

We have briefly examined the three types of love described by the ancient Greeks, four modern philosophical conceptions of the same, and what neurobiology (and to a lesser extent evolutionary biology) can tell us about the subject. How are we supposed to put these apparently disparate views and pieces of information together? From a philosophical standpoint, we need to recognize that the ancient Greeks' classification was an attempt to systematize the common types of love (using the word broadly) that characterize the human experience, while the modern philosophical discussion is usually framed in terms of what may (or may not) *justify* our loving an individual. The two are therefore not mutually incompatible. For instance, we can recognize that there is such a thing as a distinction between philia and agape and still ask whether either or both are best understood in terms of valuing the object of our love as an example of strong union, robust concern, or a particular type of emotion.

The philosophy is also not in contradiction with the science: more than was ever possible for the ancient Greeks, we now understand, for instance, both the neurological underpinnings and the evolutionary origins of eros. The modern philosophical theory of love as emotion can be enhanced and informed by what biologists tell us concerning the reason for the existence of human emotions in general and the emotions accompanying love in particular, as well as what they say about the mechanisms of emotion. Shakespeare famously asked (in *As You Like It*): "What 'tis to Love?" Neither biology nor philosophy will ever be able to substitute for the first-person experience of feeling what it is like to be in love, but they certainly give us plenty of ideas and empirical evidence to begin to answer the Bard's question in a broader sense—and hence to use our new knowledge to further enhance our enjoyment of a purposeful life.

CHAPTER 12

FRIENDSHIP AND THE MEANING OF LIFE

Of all the things which wisdom provides to make life entirely happy, much the greatest is the possession of friendship.

—Epicurus

THERE SEEM TO BE THREE MAJOR FACTORS AFFECTING our happiness: our "set point," our circumstances in life, and what we actively do about it. Or at least that's the view emerging from modern cognitive science research on happiness. The idea of a set point for happiness—that is, a stable emotional equilibrium (low, high, or whatever) to which you return quickly no matter what your circumstances are—is still controversial; indeed, there is evidence that one's set point, though it is likely to be influenced by genetic and early developmental factors, may in fact vary substantially throughout life. Regardless, data show that about 50 percent of the difference in happiness among people is due to their (quasi) set points, another 10 percent or so to the circumstances they find themselves in, and the remaining 40 percent to what they actively do to increase their happiness. Which means that hoping for a sudden inheritance (a change in circumstances) isn't as good a bet to improve your happiness as getting out and making some friends (active behavior).

Friendship is recognized by philosophers and cognitive scientists alike as one of the fundamental ingredients of human happiness. Psychologist Robert B. Hays defined friendship as a "voluntary interdependence between two persons over time, which is intended to facilitate socio-emotional goals of the participants, and may involve varying types and degrees of companionship, intimacy, affection and mutual assistance." Okay, that sounds a bit dry, but it is a concept that has prompted cognitive scientists to look into exactly how having friends improves your eudaimonia (which, you may remember, is the philosophical term for a good and fulfilling life). Their findings are intriguing.

To begin with, happiness is influenced, as one might expect, by all of the "big five" personality traits: agreeableness, conscientiousness, extraversion, neuroticism, and openness. Equally unsurprising perhaps, neuroticism gets you down while extroversion is a particularly good predictor that you'll be happy. As research conducted by Meliksah Demir and Lesley Weitekamp also clearly shows, however, friendship augments happiness above and beyond the basic effect of personality. (What is equally interesting—and may go against common lore—is that gender has no effect: men and women tend to be equally happy or unhappy as a function of their personality and friendships, with no apparent contribution from the fact that they are men or women.)

You may not be surprised to know—but it's nice to have the science to show your skeptical acquaintances—that the effect of friendship on happiness has nothing to do with how many friends you have (so much for Facebook and the like, though more on that later) and everything to do with the perceived quality of your relationships. In particular, what makes for a good happiness-enhancing friendship is the degree of companionship (when you do things together with your friends) and of self-validation (when your friends reassure you that you are a good, worthy individual). Everything else seems to be icing on the cake, so to speak.

So far in this account the science of friendship may appear a bit underwhelming: we see that it simply confirms and quantifies what most of us probably have intuited about the value of friendship. But the power of science is that it can also surprise us with findings that initially seem puzzling and even counterintuitive. For instance, controversial research carried out by James Fowler and Nicholas Christakis seems to indicate that a number of human behavioral traits "spread," following the same dynamics characteristic of infectious diseases, but through the vehicle of the ties of friendship, not viruses and bacteria. For instance, if you become obese at some point in your life, your close friends will have—quite irrespective of other factors—a whopping 57 percent chance of doing the same. Moreover, even friends of your friends will be affected, though at a lower rate of 20 percent, and *their* friends too (meaning, two degrees of separation from you) will see their chances of becoming obese going up, by about 10 percent. Feel the pressure now to cut out the doughnuts and go to the gym?

Fowler and Christakis have shown that the same is true for smoking (if you quit, your friends have a 67 percent chance of quitting too, and their friends a 36 percent chance), alcoholism, and depression, and the effect is even present for happiness (meaning subjective well-being) itself! These are rather extraordinary findings, and the authors have predictably encountered a fair share of skepticism in the scientific community. (As we saw in Chapter 8, that's just the way science works.) But major alternative hypotheses to explain their results seem likely to be ruled out. For instance, one possible explanation is that people don't cause changes in their friends but that they simply tend to associate with like people (smokers with smokers, alcoholics with alcoholics, and so on). But Fowler and Christakis have documented these changes happening over time—again, like the spreading pattern of an infection—and not simultaneously, as this explanation would predict. Another possibility that has been advanced is that we share similar environments, and those environments tend to have similar impacts on our behaviors. (For example, we are more likely to become obese if we live in a neighborhood where most of the restaurants are fast-food chains.) But again, this is an explanation that doesn't actually account for the data: Fowler and Christakis have shown that the effects hold for friends but not for neighbors, with whom we presumably share much of the same environment; in fact, we are much more likely to share an environment with our neighbors than with our friends, who, after all, may be living in a different part of town. Like it or not, our behavior makes us partially responsible—for better or for worse—for the happiness of our friends, and that presents us with even more of an ethical duty to do the right thing. Which brings us to what philosophers have to say about friendship and happiness.

Every discussion of the philosophy of friendship has to go back to the ancient Greeks, and in particular to Aristotle. We have seen (in Chapter 11) that they recognized three types of love: agape, eros, and philia. You may recall that agape is a broad kind of love, the kind that religious people feel that

God has for us, or that a secular person may have for humanity at large. Eros, naturally, is more concerned with the type of love we have for sexual partners, though the Greeks meant it more broadly than we do. Philia is the type of love that concerns us here because it includes the sort of feelings we have for friends, family, and even business partners. (It also includes nationalistic love for one's country, but that's a different story.)

At this point an obvious question arises: what exactly is the difference between love (as in eros) and friendship (as in philia)? The obvious answer is that typically (though certainly not necessarily) you have sex with your eros partner but not with your philia friends. More subtly, however, philosophers have pointed out that love is an evaluative attitude, while friendship is a relational one. It makes perfect sense that you could be in love with someone who doesn't reciprocate your feeling, but it is incoherent to say that one has a nonreciprocal friendship.

At any rate, when it comes specifically to friendship, Aristotle recognized three types: friendships of pleasure, of utility, and of virtue. In friendships of pleasure, you and another person are friends because of the direct pleasure your friendship brings—for instance, you like and befriend people who are good conversationalists, or with whom you can go to concerts, and so on. Friendships of utility are those in which you gain a tangible benefit, either economic or political, from the relationship. Exploitation of other people is not necessarily implied by the idea of utility friendships—first, because the advantage can be reciprocal, and second, because a business or political relation doesn't preclude having genuine feelings of affection for each other. For Aristotle, however, the highest kind of friendship was one of virtue: you are friends with someone because of the kind of person he is, that is, because of his virtues (understood in the ancient Greek sense of virtue ethics, as discussed in Chapter 5, not in the much more narrow modern sense, which is largely derived from the influence of Christianity).

There is a problem, however, with Aristotle's understanding of the friendship of virtue, and it's one that seems to apply also to the modern everyday concept of friendship: if we enter into a friendship with someone because we admire her character, then presumably we could meet someone else who exemplifies even better what it is that we admire in her, and we would therefore be perfectly justified in "trading up." Philosophers refer to this situation as "fungibility." We have already encountered this problem when we talked about love: presumably I love someone because of a number of traits she has, both physical traits and character traits. (Physical traits usually don't enter into friendships, which in some important sense simplifies things.) But that being the case, then I would be perfectly rational if I switched to another woman—should I encounter her (and should she be interested)—if she had the same characteristics as my current love, only better.

Although undoubtedly these scenarios do happen, both in the case of friendships and in the case of love relationships, we might frown upon someone who so callously thinks of human beings as fungible objects. One way some philosophers get around this is to point out that we often initially fall in love with a person, or become friends with a person, because of some of their characteristics, but that a sustained feeling of love or friendship is then built on our shared history as individuals, and clearly the longer our history is with a particular friend or lover, the deeper and more unique the relationship becomes. This "traits plus history" model may or may not solve the problem of fungibility, depending on whether you find it convincing. As a matter of empirical fact, some people don't (assuming they ever stop to reflect about the meaning of their relationships), since trading up is a common sociological occurrence.

Some features of friendship are obvious—such as caring for the other person, feeling and displaying sympathy, or engaging in actions that help your friend or prevent harm to him. However, it is hard to come up with a feature that is specific to friendship and friendship only. For instance, intimacy is a characteristic of friendship: we share details of our lives with our true friends that we don't share with superficial acquaintances and often don't even share with our eros companion. Then again, we may share even more intimate details, at least of a certain kind, with a therapist, who is not our friend in the standard sense of the term. (The relation with a therapist is a business relation, so for Aristotle a therapist could at most be a friend of utility.)

Why is friendship important, other than for the beneficial effects that science, as we saw early in the chapter, is beginning to demonstrate? Aristotle's opinion was that friends hold a mirror up to each other; through that mirror they can see each other in ways that would not otherwise be accessible to them, and it is this (reciprocal) mirroring that helps them improve themselves as persons. Friends, then, share a similar concept of eudaimonia and help each other achieve it. So it is not just that friends are instrumentally good because they enrich our lives, but that they are an integral part of what it means to live the good life, according to Aristotle and other ancient Greek philosophers (like Epicurus). Of course, another reason to value the idea of friendship is its social dimension. In the words of philosopher Elizabeth Telfer, friendship provides "a degree and kind of consideration for others' welfare which cannot exist outside it."

There is one more intriguing philosophical aspect of the concept of friendship. We saw in Chapter 5 that there are three major theories of ethics: virtue ethics (originally from Aristotle), consequentialism (or more narrowly, utilitarianism, rooted in the philosophies of Jeremy Bentham and John Stuart Mill), and deontology (particularly à la Immanuel Kant). I argued during that earlier discussion that one could take interesting elements from each of these approaches to build a more personalized and flexible view of ethics. Some philosophers, however, have suggested that friendship as we understand it may naturally fit with Aristotle-type virtue ethics (not surprisingly, given all he had to say about friendship!), but is difficult to reconcile with both consequentialism and deontology. How is this possible, and what does that imply?

The idea is that a major difference between virtue ethics, on the one hand, and consequentialism and deontology, on the other, is that virtue ethics is an individually centered concept of ethics: the fulcrum of moral reasoning is the person, her character traits, and how she relates to people around her. Consequentialism and deontology, on the contrary, are agent-neutral: they attempt to establish universal criteria of ethical behavior by which all individuals are treated in the same way (for example, Mill's idea of maximizing pleasure and minimizing pain, or Kant's categorical imperative of engaging only in behaviors that one could accept if they became universal laws). The problem, then, is that friendship is *by definition* a type of relationship in which we have a distinctive *moral* preference for a particular person. (This holds also for love of the eros type, as well as for the type of philia we have for members of our own family.) For instance, loyalty to a friend, helping her in ways that might conflict with our general duty toward other human beings, is something that Aristotle would not have found at all problematic (within limits, of course: one is still not allowed, say, to steal or kill for a friend). Mill and Kant, on the other hand, would be somewhat at a loss if asked to justify the special regard one has for a friend within the context of their broader ethical systems. I'm not sure this argument is sufficient to embrace virtue ethics over its rival moral theories, but it is something to ponder while you carve your own ethical system out of the options made available to us by philosophical reasoning.

These days we simply cannot have a chapter on friendship without briefly discussing the early-twenty-first century phenomenon of virtual friends and social networks. (Full disclosure: I do have both an "official" Google Plus page and a Twitter feed, @mpigliucci, not to mention a more "private" Facebook page that I reserve for actual friends and family.) Let's start with a 2007 article in *The New Atlantis*: Christine Rosen wrote that "the Delphic oracle's guidance was know thyself. Today, in the world of online social networks, the oracle's advice might be show thyself." As much as this quip might be pleasing to any curmudgeon with philosophical inclinations and other assorted Luddites, I think it is far too much of a caricature of what is actually going on with social networking.

There is no doubt that social networking is a phenomenon of global and staggering proportions: by January 2011, Facebook, which had launched only in February 2004, counted about 600 million users, out of a total of about 2 billion people using the Internet worldwide. Surely this level of participation reflects a powerful urge that people have to present themselves to the world, to communicate, and yes, to engage in (mostly harmless) social exhibitionism. But is this phenomenon, as Slate critic Michael Kinsley put it, a "vast celebration of solipsism"? Again, that comment seems far too quick and dismissive. Both Rosen and Kinsley hastily dispatch of something that is simply a modern incarnation of an ancient yearning: humanity has always had a need to share social information (often dismissively referred to as "gossip"). After all, what distinguishes us from other primates is not that we are social (so are chimpanzees and macaques), nor that we communicate (so do birds, social insects, and countless other species). It is that we share complex bits of information through language. Social networking in the Internet age has made that possible to a degree that would have been inconceivable even a few years ago; although social networking has generated its own peculiar dynamics, the basic urge remains the same.

In 1967 psychologist Stanley Milgram carried out a famous experiment whose results have become standard social lore in the guise of "six degrees of separation." He investigated how many connections would be necessary for a chain letter sent within the United States to reach a prespecified target individual. On average, that number was 5.5. The experiment was repeated in the era of email by Duncan J. Watts at Columbia University and expanded to worldwide targets. The result was about the same: five to seven steps. Interestingly, however, what both Milgram and Watts demonstrated was not that we are "six friends removed" from, say, Kevin Bacon, but that the most efficient way to reach prespecified targets is through loose acquaintances, what social scientists call "weak ties," rather than through friends. Accordingly, the sort of information that travels best via social networks is the same kind that is spread well via weak ties: fads, gossip, rumors, but also links to interesting newspaper articles, broadcasts, book reviews, movies, events, or opinions.

These results clearly point to why exaggerated criticisms of—or for that matter, exaggerated optimism about—social networking are, well, exaggerated, and probably grossly so. Facebook, Twitter, Google Plus, and the like are hardly going to spell the end of civilization as we know it, no more so than television, movies, or even the printing press itself did before them. (And of course, fearmongers and doomsayers were present and very vocal on all those other occasions as well.) But social networking by itself is certainly also not going to start revolutions, solve climate change, or significantly ameliorate any of the many real problems facing humanity. What these new tools have been doing is simply help us keep up with friends and families who don't live in our same town, potentially introduce us to a wider circle of people than we might otherwise have had access to, and make it possible for a number of interesting news items to cross our screen that we might have missed on our own. All this does not amount to an insignificant contribution to our lives, but it also doesn't radically alter who we are, how we think, or most important, how we act.

Another thing social networking isn't going to do is get you friends, no matter how long your "friends" list may be. As Christine Rosen put it in her article, "The idea of public friendship is an oxymoron," and Aristotle would have certainly agreed. One simply *cannot* have hundreds or thousands of friends, because friendship is built on trust and intimacy and requires a significant investment of time. Moreover, friends cannot (or should not) be "managed" or "edited," or whatever the going lingo of social networks happens to be.

Which is why, the next time you are having dinner out with an actual friend, you really should do her a favor and refrain from checking on your virtual acquaintances until you are back in the privacy of your home. If someone is a friend, the least she deserves from you is your undivided attention for a few hours a week.

PART V
THE (POLITICAL) ANIMAL INSIDE YOU

CHAPTER 13

RIGHT, LEFT, UP, DOWN: ON POLITICS

> Politics is the art of looking for trouble, finding it, misdiagnosing it, and then misapplying the wrong remedies.
> —Groucho Marx

Our old friend Aristotle famously said that "man by nature is a political animal." According to modern science, we are not the only political animal, at least not for primatologist Frans de Waal, author of *Chimpanzee Politics: Power and Sex Among Apes*. From Aristotle to de Waal, however, philosophers and scientists alike have agreed that politics is fundamental for our social life, and therefore politics is a topic we need to examine in our quest for the intelligent person's guide to a flourishing existence.

Philosophers and political scientists tend to think of politics in rational terms, considering how people might react to arguments and counterarguments about this or that political position or ideology. As it turns out, however, much that goes on inside our minds when we think about politics has precious little to do with rationality and a lot to do with personality. Accordingly, we are going to start with some recent findings about human politics garnered from the biological and cognitive sciences, just to get us in the mood.

For instance, Douglas Oxley and his collaborators have shown that there is a link between people's political attitudes and—of all things—their physiological traits. Let me explain. The researchers asked a number of subjects questions concerning their political opinions on eighteen topics and then divided the respondents into two groups depending on whether they exhibited a high or low preference for "protective policies"—that is, whether they favored a range of positions about immigration, gun control, defense, foreign aid, and so on, that are typically associated with a conservative (high preference) or a liberal (low preference) point of view. They then measured the same subjects' skin conductance in reaction to threatening images, as well as their startle response in reaction to sudden loud noises. Both skin conductance and startle responses are physiological measures of an individual's emotional reaction to threat, and they cannot be controlled consciously.

The results were, well, somewhat startling. Simply put, conservatives—but not liberals—had a strong response to both threatening images and startling noises, which means that these people were *physiologically*—not just ideologically or rationally—more sensitive to threats. Think about the implication here: people may arrive at certain general political ideologies, not through deliberation, but because of fundamental aspects of their personal psychology (however those developed). And there is more: in what can be seen as a follow-up study, a different group of researchers headed by Ryota Kanai more recently published the finding that something like these same psychological differences can be clearly seen at the level of gross brain anatomy. As it turns out, liberals tend to have an enlarged anterior cingulate cortex, while conservatives are big in the amygdala. What does this mean? Well, the anterior cingulate cortex is involved in the ability to deal with conflicting information, while the amygdala handles response to threats. You can draw your conclusions from this research, but I warn you that throughout the rest of this chapter you will be presented with scientific evidence that backs up many common stereotypes about liberals and conservatives (right down to what kind of objects they prefer to have in their bedrooms!)

Of course, one conclusion we should *not* draw from these findings is that there is any such thing as a conservative (or liberal) gene that makes people think in a certain way, evidence and reason be damned. Things are much more complicated than that—fortunately. Indeed, in some respects it is not at all surprising that differences in behavior (like political opinions, or responses to threat) can be traced back to the human brain. Where else would we find them? And just because there are *biological* differences between brains doesn't mean that those differences are *genetically* based. (It is a common mistake to confuse biology with genetics: pretty much everything human beings do is biological, but not all of it has to do with genetics.) And even if researchers discover some genetic bases for biological differences between brains (and there are some hints that they may), they almost surely would not come down to a small number of genes having direct and irrevocable effects on complex behaviors such as political opinions. Still, it begins to look like politics is another one of those fundamental human characteristics (morality being another; see Chapters 3 and 4) that traces its roots straight down to what kind of biological beings we happen to be.

As interesting as science-based analyses of the conservative-liberal dichotomy may be, they often buy into what an increasing number of political scientists see as a far too simplistic view of the political landscape. Indeed, research shows that the 1990s and 2000s in the United States have seen an odd pair of trends develop: on the one hand, the two major parties have become increasingly polarized in their positions, and on the other hand, the American people have, if anything, developed more of a consensus on a variety of issues that usually mark political elections. A study by Delia Baldassarri and Andrew Gelman found, among other things, that half of Republicans didn't think of themselves as conservative, and only 12 percent of the interviewed considered themselves simultaneously Republican, conservative, and opposed to abortion. Findings are similar among Democrats (only 36 percent of whom called themselves liberal), a whopping 90 percent of whom espouse nonliberal views on abortion, aid to black people, and government intervention. Clearly something is amiss in the simplistic picture of the American political landscape we all get from the news media.

The data suggest a more subtle and intriguing picture: what has apparently happened since the great political realignment of the 1960s and 1970s (when the Democrats "lost" the South because of their support for civil rights) is that the parties themselves have been better at "sorting" (to use Baldassarri and Gelman's term) voters—that is, at channeling people to vote one way or the other despite a demonstrably low correlation between people's political opinions and party platforms. The research in question, for instance, showed that the correlation between party affiliation and political issues has increased by 0.05 per decade in recent times, making the two political parties more easily identifiable as proxies for whatever issue a person cares about. (For example, if you are "pro-life" on the issue of abortion, you increasingly realize that you need to vote Republican, while if you are "pro-choice" you know that your best bet is with the Democrats.) But what political scientists also found was that correlations between issues themselves have hardly moved at all during the same period, so that increased partisanship has not been accompanied by increased ideological alignment—contrary to popular opinion fostered by the media's simplistic narrative.

The same author, Delia Baldassarri, and her collaborator Amir Goldberg published a follow-up in which they investigated the nature of political belief networks; again, they were seeking evidence of how people think politically, instead of just buying into the simplified conservative-liberal bipolarism. What they found was at the very least evidence for a tripolarism, which sheds some light onto the otherwise increasingly puzzling political discourse in the United States (and, I suspect, in other Western countries as well). The tripolarism results from Baldassarri and Goldberg's discovery that people, broadly speaking, fall into three categories when it comes to political attitudes. What they call "ideologues" are the sort of people who readily identify themselves as either conservative or liberal in the very same way in which we usually think of those two categories. (I am a liberal "ideologue" in this sense, and accordingly I tend to favor solid social welfare programs to protect the poor and elderly, as well as an expansion of civil rights to protect both traditional and less traditional minorities, such as gays.) The second group, which the authors label as "alternatives," are not quite that easy to define because they disassociate moral and economic attitudes.

(So, for instance, they may be socially progressive but economically conservative, or the other way around.) Finally, the third group is labeled "agnostic" because its members tend to show very little integration of their different political beliefs.

Baldassarri and Goldberg then investigated the effect that individual social identities play in making someone an ideologue, an alternative, or an agnostic. Again, the results did not align with the standard narrative. For instance, it turns out that people with high income tend to be morally conservative if they belong to the ideologue group, but they are morally liberal if they are among the alternatives. Moreover, people with weak religious commitment are alternatives if they are also high-earners, but so are people who have strong religious commitments but are low-earners. To complicate things further, ideologues who are either highly or poorly educated (but not in the middle) tend to identify with Democrats, while among alternatives more education signals a higher likelihood of being a Republican. Does your head spin now? Can you see why such findings wouldn't nicely fit into a two-minute sound bite on the evening news? And yet, these results mean that if we are serious about understanding political opinion, we have to start acknowledging that there is no simple pattern connecting someone's ideology to their income or religiosity. Oh, and here is the kicker: other things being equal, when people are faced with having to choose among sets of positions that are not exactly aligned with their own preferences, they tend to go conservative, which might give the Republican Party (and probably other conservative parties in the Western world) a built-in advantage. This advantage would explain what many political scientists have always considered a paradox: most Americans agree significantly more with Democrats on a number of issues, and yet tend to vote more for Republicans.

If there is more ideological variety among people than one might expect from a two-party system, how is it that the parties in question have managed to polarize the debate nonetheless? One way is through an approach that we have encountered already in a different context in Chapter 6 and that every intelligent citizen ought to be aware of: framing. People respond differently to the same factual information depending solely on how it is presented, or "framed." For instance, it makes a difference whether your doctor tells you that a delicate operation you need to undergo has an 80 percent survival rate or a 20 percent mortality rate, even though factually the two scenarios are identical. Politicians and advertising agencies know this very well, which is why they try to prevent negative connotations for any of their positions or products. (To go back to our earlier example, some people say that they are "pro-life," not "anti–reproductive rights," others say that they are "pro-choice" rather than "anti–fetus rights," and so on.)

According to political psychologist Rune Slothuus, framing—within the political context—can work through two different mechanisms: by affecting people's considerations of the importance of certain factors in a given debate, or by changing the content of the discussion itself. In the first case, the politician is trying to focus your attention on aspects of the problem that are more likely to change your mind about the issue at hand (and, conversely, trying to deemphasize other aspects concerning which you may disagree with him); by contrast, content framing attempts to introduce new arguments that the target audience may not have considered pertinent to the discussion or may not have been aware of. The empirical evidence is that both types of framing occur, but they work differently with different people—a fact that you may want to be aware of the next time you need to make up your mind about how to vote on an issue (or, for that matter, whether to buy a given product).

In particular, Slothuus maintains, different types of framing are mediated by an individual's level of political awareness and the strength of his or her political values. (The two are clearly not the same: people may have strong political opinions despite not being especially conversant in the specifics of an issue, and vice versa—people may know a lot about an issue and yet not endorse a strong partisan position on it.) To test this idea, Slothuus exposed his experimental subjects to different versions of a newspaper article on an actual pending social welfare bill that included giving incentives to unemployed people to get back into the workforce. One version of the article framed the issue in terms of the jobs that would be created as a result of the bill, and another version framed it in terms of the increased hardship to the poor that would result from the bill. Both the importance of various aspects of the discussion and its content in terms of arguments pro and con were changed to test Slothuus's ideas about the different types of framing.

The results were spectacularly clear: framing did make a difference, in the sense that the bill received higher support when presented as an opportunity to create jobs than when portrayed as a source of hardship for the poor. (Remember, again, that the factual information was the same across the different versions of the newspaper article, so people were responding to the framing, not the facts.) "Importance framing" (emphasizing or deemphasizing certain arguments) affected how moderately and highly politically aware people reacted, but had no effect on people with low political awareness. By contrast, "content framing" (bringing in new arguments that might have been missed by the audience) affected the moderately politically aware, but did nothing for both those with heightened political awareness and those whose political awareness was very low. Notice not only that framing works through different psychological mechanisms, but that people with low political awareness are quite resistant to framing—presumably because they know too little about what is going on to appreciate subtle shifts of emphasis or content. What about the effect of framing on people of varying strengths of political commitment? Here, perhaps not surprisingly, framing did not have an effect on the highly politically committed, but did sway people with lower levels of commitment. Apparently, you can change the mind of either moderately to highly informed people or those who are not yet married to a given ideology.

If you are still not discouraged about the chances of rational political discourse in our society, there is another study that very likely will do the job: the research of Monica Prasad and her colleagues into cognitive dissonance and political opinion. (Of course, if you are naturally optimistic, like me, then you will simply see what follows as useful knowledge that will help you guard yourself against the sort of rationalizing that we are about to encounter. Take your pick.) We already came across the concept of cognitive dissonance in Chapter 6, when we were discovering that another of Aristotle's famous characterizations of humanity—the rational animal—wasn't quite as true as the Greek sage might have thought or hoped. There, however, we were talking about people who literally have a split brain and who have helped neurobiologists show what is going on inside our heads when we make up stories to make sense of reality. The subjects in Prasad's study, instead, were perfectly normal people who just happened to be convinced—against all evidence to the contrary—that there was a link between the former Iraqi dictator Saddam Hussein and the terrorist attacks of September 11, 2001, on American soil. (As a disclaimer, I should note that there are some methodological problems and limitations with Prasad's study, some of which are discussed in the paper itself, so put on your critical thinking cap and consider what I'm about to discuss as only tentative.)

The researchers focused on this particular politically based belief not only because, as they put it, "unlike many political issues, there is a correct answer," but also because the belief was still held by about 50 percent of Americans as late as 2003—despite the fact that President Bush himself at one point declared that "this administration never said that the 9/11 attacks were orchestrated between Saddam and Al Qaeda." Prasad and her colleagues didn't set out to pick on Republicans, by the way; they wrote that they would have fully expected to find similar results had they conducted the study a decade earlier and targeted Democratic voters' beliefs about the Clinton-Lewinsky scandal.

The hypotheses tested in this study were two alternative explanations for why people hold on to demonstrably false political beliefs. The "Bayesian updater" theory says that people change their beliefs in accordance with the available evidence; therefore, a large number of people held on to the false belief of a connection between Hussein and 9/11 because of a subtle, concerted campaign of misinformation by the Bush administration (despite President Bush's own statement that there was no such connection). (The term *Bayesian* refers to a type of probability theory that has become popular in both scientific and philosophical circles and is named after Reverend Thomas Bayes, who proposed it in a short and later very influential article originally published in 1763.)

The alternative theory tested by Prasad's group is what they call "motivated reasoning": the battery of cognitive strategies that people deploy to avoid facing the fact that one of their important beliefs turns out to be factually wrong. The results of this study are illuminating well beyond the specific issue of Hussein and 9/11: the same strategies are used even by well-informed and well-educated people in a variety of circumstances, from the political arena to the defense of pseudoscientific notions such as the alleged (and nonexistent) link between vaccines and autism. Indeed, it may be a good learning experience to reflect on whether you have recently been guilty of deploying such cognitive shields in defense of your own cherished (and possibly indefensible) notions. You may want to ask your friends about this, since they are more likely to hold up a truthful mirror than the one you might use on yourself (see our discussion of friendship in Chapter 12).

The first thing that Prasad and her colleagues found was that, not surprisingly, belief does translate into voting patterns: interviewees who answered the question correctly—that is, who knew that there was no demonstrated connection between Saddam Hussein and the 9/11 attacks—were significantly less likely to vote for George Bush and more likely to vote for John Kerry during the 2004 presidential election.

Of the remainder, how many behaved like Bayesian updaters, changing their opinion on the matter once presented with evidence (namely, President Bush's own speech) that there was no connection? A dismal 2 percent. The rest of those who stuck with their original opinion, evidence to the contrary be damned, deployed a whopping six different defensive strategies, which Prasad and her colleagues characterized in detail. Here they are, in decreasing order of importance:

- *Attitude bolstering* (33 percent): Or, as Groucho Marx famously put it, these are my principles, if you don't like them, I've got others. This group simply switched to other reasons for why the United States invaded Iraq, implicitly granting the lack of a Hussein-9/11 connection and yet not being moved to change their position on the actual issue, the Iraq War.
- *Disputing rationality* (16 percent): As one of the interviewees put it, "I still have my opinions," meaning that opinions can be held without or even against evidence, simply because it is one's right to do so. (Indeed, one does have the legal right to hold on to wrong opinions under American law, as it should be; whether doing so is a good idea is an altogether different matter, of course.)
- *Inferred justification* (16 percent): "If Bush thinks he did, then he did it." The reasoning here is that there simply must have been a reason for the good guys (the United States) to engage in something so wasteful of human life and resources as a war. The fact that they couldn't come up with what exactly that reason might have been did not seem to bother these people very much.
- *Denial of belief in the link* (14 percent): These were the subjects who had said they believed in the link between Iraq and 9/11 but who, when challenged, changed their story, often attempting to modify the original statement, as in: "Oh, I meant there was

a link between Afghanistan [instead of Iraq] and 9/11."
- *Counterarguing* (12 percent): This group admitted that there was no direct evidence linking Saddam Hussein and the terrorist attacks, but nevertheless thought that it was "reasonable" to believe in a link, based on other issues, such as Hussein's general antipathy for the United States, or his "well-known" support of terrorism in general.
- *Selective exposure* (6 percent): Finally, there are people who simply refuse to engage the debate (while not changing their mind), adducing excuses along the lines of, "I don't know enough about it" (which may very well be true, but of course would be more consistent with agnosticism on the issue).

How are any of these responses possible? Are people really so obtuse that they will largely fail to behave as "Bayesian updaters" by taking the rational approach to assessing evidence and belief, even when the evidence is readily available? There is no need to be too hard on our fellow humans—or indeed on ourselves, since all of us are likely to behave in a very similar fashion in a variety of circumstances. What is going on here is that most of us, most of the time, use what cognitive scientists call "heuristics"—convenient shortcuts or rules of thumb—to quickly assess a situation or a claim. There is good reason to do this, since on most occasions we simply do not have the time and resources to devote to serious research on a particular subject matter, even in the Internet and smart phone era of information at our fingertips. Besides, sometimes we are simply not sufficiently motivated to do the research even if we do have the time—the issue might not be important enough to us compared to our need to shop for groceries or get the car cleaned.

Unfortunately, it is also heuristically efficient to stand by your original conclusion once you reach it, no matter how flimsy the evidence you considered before reaching it. Again, this is simply a matter of saving time and energy. As a result, we use politicians we trust, political parties, or even celebrities as proxies to make up our minds about everything from the war on Iraq to climate change science, and once we adopt a position we deploy our cognitive faculties, if challenged, toward deflecting criticism rather than engaging it seriously. This was demonstrated on a variety of occasions well before the Prasad study. For instance, following the heuristic, "if someone who seems to know what he is talking about asks me about X, then X is likely to exist and I should have an opinion about it," people volunteer "opinions" (that is, they make up stuff out of thin air) concerning legislation that does not exist, politicians who do not exist, and even places that do not exist! Which brings us back to the study on Hussein and 9/11: as we have seen, many people apparently used the heuristic, "if we went to war against country X, then country X must have done something really bad to us"—in other words, there *must* be a reason!

There is serious doubt whether humans are the rational animal (though by all means we are more rational than every other species, in the sense of having the ability to think about what we do and its consequences), and we may not be the only political animals either, considering the mounting evidence on the politics of chimpanzees and other social primates. Nonetheless, both politics and rationality play a very important part in defining what it means to be human, and hence in giving meaning to our existence. The next time you find yourself vehemently defending a political position, however, you may want to seriously ponder whether you are behaving as a Bayesian updater or whether—more likely—you are deploying one of the six rationalizing strategies highlighted earlier. If so, your internal Bayesian calculator may require some tuning up. It would be good for you, and it would be good for our society.

CHAPTER 14

OUR INNATE SENSE OF FAIRNESS

> Fairness is what justice really is.
> —Potter Stewart, US Supreme Court judge

If you have trouble sleeping or adjusting to jet lag, you may want to try a chemical known as 5-hydroxytryptamine, more commonly known as serotonin. Then again, you may want to be careful, because it turns out that serotonin directly affects your propensity to judge situations in a fair way—a crucial component of our social and personal life. Research conducted by Molly Crockett and her colleagues at the University of Cambridge and the University of California at Los Angeles shows that lower levels of serotonin in the brain cause people to be more intransigent about what they consider their unfair treatment at the hands of others.

The researchers used a variant of the so-called ultimatum game, in which subjects were asked to accept or reject a given offer to split a sum of money. Most people consider a 45/55 percent split, even if they get the short end of the stick, to be within the boundaries of fairness. When the split shifts to 30/70, subjects say the deal is unfair, and they think of it as highly unfair when the split is about 20/80. What Crockett and her collaborators did was to submit some of their subjects to an acute tryptophan depletion procedure, which interferes with the production of serotonin, temporarily lowering it significantly. To make sure that their results were not biased by the subjects' knowledge of whether they had received the procedure or not, a placebo control was also used during the experiment. Moreover, to achieve the highest standard of scientific investigation, the whole thing was done using a double-blind protocol: because the scientists analyzing the data did not know which subjects had received the treatment and which had received the placebo, any possibility of unconscious bias in the interpretation of the results was eliminated.

The outcome was pretty clear-cut: there was no difference between treated and placebo subjects when the offer was fair, or even somewhat unfair; when the offer was unquestionably unfair, however (the 20/80 proposal), a much larger percentage of serotonin-depleted subjects rejected the offer when compared to the controls. Less serotonin circulating in your brain automatically (that is, without your conscious realization of it) lowers your threshold of tolerance for unfair treatment. (It is possible that more serotonin will make you temporarily more tolerant of unfairness, but the study conducted by Crockett and her colleagues did not address this possibility.) What makes this research particularly interesting is that it is not the only piece of neurobiological information we have about how the brain weighs fairness: it turns out that a similar behavior is observed in patients with lesions to the ventral prefrontal cortex. Researchers interpret this to mean that serotonin is probably achieving its effect through a modulation of the activity in the ventral prefrontal cortex. But that is not all: we also know that interference with a nearby brain area called the dorsolateral prefrontal cortex—for instance, by disrupting its activity using transcranial magnetic stimulation—achieves the opposite effect, making people more likely to accept unfair offers. It's like the brain has these two regions that join up to weigh a reaction to the potential instances of unfairness in which we find ourselves. These findings are quite stunning and indicate that "fairness" is not just a cultural construct or a matter for theoretical philosophical discussions; in fact, we literally have a fairness calculator embedded in our brains!

Not surprisingly, the concept of fairness itself is a matter for serious discussion among philosophers, and as we shall see, some philosophical treatments of the matter mesh with the science significantly better than others—the sci-phi interface once again providing us with the best available picture on which to base our views and make our decisions.

Let us start with a very important concept in ethical and moral philosophy, the idea of "reflective equilibrium," originally introduced by Nelson Goodman in 1955 (though he didn't use this term) and eventually made famous by one of the most influential moral philosophers of all time, John Rawls. In essence, the method of reflective equilibrium, as the name implies, is a type of rational reflection that seeks to achieve an equilibrium among different notions, judgments, or intuitions we might have about a given ethical problem (or any other problem, for that matter). The goal is to continuously revise our judgments and reasons until they become as coherent as possible, thus allowing us to achieve said "equilibrium." For instance, I may hold a starting position that abortion is to be avoided because it amounts to the extinction of a potential human life. However, I also think that a woman should have as much control as possible over her own reproductive fate. Moreover, I think that the welfare of the mother should override that of the fetus when they are in conflict, because the first is a fully formed person with rights and the latter is a potential person with a more limited range of rights. A process of reflective equilibrium would force me to acknowledge all these different moral intuitions, work through why exactly I hold them, and highlight when they are in conflict with each other. I would then proceed (probably with the help of a friend who would function as a sounding board for my complex and perhaps contradictory thoughts on the matter) to reconcile as many of my different intuitions on abortion as possible, and even to consciously reject or modify some of those intuitions once I see more clearly what they consist of.

We shall see in the next chapter exactly how Rawls applied the method of reflective equilibrium to arrive at an ingenious device meant to guide us toward the establishment of a society that is as just as can be rationally conceived. For now, however, let us note that some of the other philosophical positions on ethics that we have encountered in our previous discussions are clearly at odds with the idea of reflective equilibrium—as sensible as the latter appears to be at first glance. For instance, utilitarians, who think we should maximize happiness for the greatest possible part of the population, have at least two problems with an approach to ethical decisions that seeks coherence among different ethical intuitions. On the one hand, they question whether we should be taking intuitions in matters of morality seriously at all: where do these intuitions come from, and why should we trust them? On the other hand, anyone seeking coherence in their ethical thinking is willing to question, revise, and possibly reject their own rules or priorities in matters of morality. Since utilitarianism is based on one such cardinal rule (the pursuit of the course of action most likely to benefit the largest number of people), obviously utilitarians feel uncomfortable with the whole idea of reflective equilibrium. The rejoinder to the second utilitarian objection is philosophical in nature, while a good response to the first objection comes from neurobiology (and, as the reader will surely have guessed, from the sort of evolutionary biology that we discussed when talking about the evolution of morality more broadly).

We will consider the philosophy first, then move back to science. As with everything else in life, we simply cannot afford the time, energy, and quite frankly the pain of continuously revising our assumptions about what we do and how we do it. It may be true, as Socrates famously said, that the unexamined life is not worth living, but it is also true that we better spend most of our time actually *living* said life. This is why proponents of the reflective equilibrium approach make a distinction between a broad and a narrow application of the principle. To take a narrow approach to reflective equilibrium is to seek balance among accepted principles and moral intuitions, without going so far as to question the origin or reliability of said principles and intuitions. For instance, going back to our brief discussion of abortion, your moral intuition may well be that life has to be protected at all costs, and this intuition becomes an underlying assumption that guides all your treatments of the issue. Even holding on to such an intuition, however, there may be significant room to achieve a reflective equilibrium among other elements of the problem—for instance, how to balance the right to life of the fetus and that of the mother. (After all, both have a right to life, so if the two are at odds we need to agree on whose right trumps the other's.)

However, from time to time we may want to expand the debate to question some of the cardinal principles, such as the idea that life has to be protected no matter what. That principle may turn out to be in contradiction with other moral positions held by an individual—for instance, someone with a strong moral intuition that makes him favor the death penalty. If life is sacred—in either a religious or even a secular sense—then it would seem that being against abortion but in favor of the death penalty generates a tension between different moral intuitions. This is where a broad conception of reflective equilibrium comes into play: we are now expanding the circle, so to speak, and considering the possibility that perhaps one (or more) of our cardinal rules or fundamental intuitions is wrong or needs to be revisited. The point is that, contrary to the claims of utilitarian critics, the idea of reflective equilibrium is compatible with the tendency we have not to question all our principles all the time. Indeed, this is what is going on when we take the broad approach to reflective equilibrium; it does not apply, however, to the narrow approach, which is certainly more frequently useful.

What about the other crucial objection to applying reflective equilibrium to ethics—that when we use "moral intuitions" of one sort or another we have no grounds to justify such usage at all because we do not know where these intuitions come from and whether they are valid? Here is where our discussion turns back toward science.

As we have seen, it turns out that we have a built-in "fairness calculator" in our brains. It appears that the calculator does not work along exclusively utilitarian lines, but rather by seeking an equilibrium (not reflective in this case, but unconscious) between different criteria. Recall (from Chapter 3) the work of Ming Hsu, Cédric Anen, and Steven Quartz, showing that three regions of the human brain are differentially engaged when we assess the fairness of a situation. The so-called putamen area seems to be in charge of assessing the efficiency of a situation (the brain's version of utilitarianism), while the insula is concerned with assessing the degree of inequity inherent in a particular decision. When the two are at odds—as they often are—a third brain circuit, the caudate-septal subgenual region, mediates between the insula and the putamen area and comes up with the final decision.

The Hsu team uncovered this by subjecting people to brain scans while they were intent on making decisions about how to best allocate funds to orphans in Uganda. Their decisions involved taking away and redistributing meals to three groups of children, and the experiment was set up in a way that the subjects had to balance their perception of inequity with their assessment of the efficiency of the resulting allocations. The overall results of the study support the idea that people tend to make moral decisions while guided principally by their sense of fairness, even at the expense of efficiency of resource distribution (within limits). In other words, our brains do not work like utilitarian calculators, but rather seem to follow a sense of, as Rawls put it, justice as fairness. Moreover, these researchers' results were congruent with the idea that this is achieved not through a method of rational deliberation but rather because of emotionally grounded moral intuitions. This finding may seem to go against the whole idea of a reflective equilibrium: after all, if we reach decisions by intuition, we don't actually *reflect* on them! Indeed, Hsu and his colleagues explicitly mention in their article (which was published in *Science*, not in a philosophy journal) that this contradicts Rawls's approach to some degree. But it doesn't really. As we have seen, reflective equilibrium has to start with some assumptions and intuitions, and when used in the narrow mode it does not have to go as far as questioning them. Indeed, the work done by these researchers offers a partial insight into the issue of where these intuitions come from: apparently, they are embedded in the inner workings of our brains. Whether this is the result of acculturation or of a long period of evolution as social animals, or both, is another matter, which, just like any other nature-nurture issue in humans, is going to be difficult to settle.

Despite the rather compelling evidence from neurobiology that our brains are wired to be "fairness calculators," the question remains of why that should be in the first place. We have discussed the possible evolutionary origins of human morality to some extent (see Chapter 4), but another piece of the puzzle that has emerged from recent research on human developmental psychology will help us get a more informed perspective on fairness and its origins. Ernst Fehr, Helen Bernhard, and Bettina Rockenbach published a paper in *Nature* in which they investigated differences in social behavior between human children of different ages in order to gain a developmental insight into our evolved differences with our closest evolutionary relative, the chimpanzee.

The study's setup was pretty straightforward. A number of children between the ages of three and eight were asked to make decisions about whether and how to share food in the form of some sweets. They could, for instance, decide either to share the candy equally with other children or to keep the candies all for themselves. The authors made sure that the recipients of the gifts were not present in the room and that the game was played only once. That way the children's decision would be a better reflection of their propensity for fairness, not of (conscious or unconscious) calculations of future expected rewards or of the need to build a reputation in the group. (Note, however, that it is impossible to control all pertinent variables in this kind of study: for instance, the children's behavior may have been affected by what they felt the adults administering the test thought of their decisions. It is really tricky to do experiments with conscious animals.)

The main result of this study was a pretty sharp distinction between the behavior of very young children (ages three to four) and that of slightly older ones (ages seven to eight): the older group had a significantly higher propensity to share their sweets, exhibiting what researchers refer to as "other-regarding" behavior. In other words, while young children tend to be self-centered, a concept of social fairness seems to emerge later on during the development of a human being. This is important because chimpanzees, for instance, do not show other-regarding preferences in their behavior and essentially always act like three- to four-year-old humans. The other interesting point to note here is that human children do engage in helping behavior when they are very young, but of an entirely different sort. Cognitive scientists refer to it as "instrumental helping": humans as young as fourteen to eighteen months readily help others achieve certain goals, such as arranging objects or opening doors, even without receiving a reward. Importantly, chimpanzees also engage in instrumental helping, strengthening the parallel between our close evolutionary relatives and the very young of our own species.

Michael Tomasello and Felix Warneken, commenting on the paper by Fehr and his colleagues, pointed out that the picture emerging from developmental and evolutionary studies is of a natural progression toward more and more sophisticated social-ethical behaviors: we start with instrumental helping (shared with chimps), move to forms of other-regarding behavior (unique to humans), and finally engage in the sort of complex reciprocal altruism that is influenced by considerations such as reputation achieved within the group, as has been repeatedly shown by studies conducted on adult humans. There is, however, an important catch: typically, human beings' other-regarding behavior is limited to members of one's group and does not extend, or is extended only reluctantly, to members of other groups. This psychological mechanism is responsible for the racism and xenophobia that underlie major conflicts between human beings, both historically and in today's increasingly multicultural society. It seems that biologically grounded social instincts only get us so far in expanding the circle of beings we are ready and willing to treat fairly (and this does not even begin to address, of course, the question of animal rights). To bootstrap our morality beyond that point, it seems that we need some more sophisticated philosophically based reflective equilibrium, which brings me to the topic of the next chapter: justice.

CHAPTER 15

ON JUSTICE

Justice is the crowning glory of the virtues.
—MARCUS TULLIUS CICERO

WHY SHOULD WE EXPECT JUSTICE IN THE WORLD? Dennis Wholey, an American television producer and author of a number of self-help books, famously said, "Expecting the world to treat you fairly because you are a good person is a little like expecting a bull not to attack you because you are a vegetarian."

The same question has been asked by philosophers ever since Plato wrote the *Republic* twenty-four centuries ago. In a famous passage in that book, Glaucon, one of the minor characters, tells Socrates about the myth of Gyges's ring, challenging the great philosopher to provide a reasonable answer to the moral of the myth. The story goes that Gyges was a shepherd in the kingdom of Lydia, in western Asia Minor (modern Turkey). One day Gyges finds a cave, and inside lies a corpse wearing a golden ring. Gyges takes the ring and discovers its magical property: by turning it he can make himself invisible at will! Needless to say, Gyges immediately puts his newfound power to work by going back to the capital, seducing the queen, killing the king, and installing himself as the new monarch of Lydia. Furthermore, he apparently got away with it, since one of his descendants is said to be King Croesus, an actual historical figure who lived in the sixth century BCE and who became synonymous with wealth.

Glaucon is rightly puzzled by the story (which may remind you of J. R. R. Tolkien's "One Ring," or—in a different form—Nicholson Baker's delightful novel *The Fermata*). He asks Socrates on what grounds one can argue that Gyges should *not* have acted as he did, given that he had the power to act and to escape punishment. It is, of course, an old conundrum: what rational argument can one produce to defend the concept of justice against the "might makes right" sort of attitude that has been all too common throughout human history? For Glaucon, the story's moral is that ethics is a social construction and therefore arbitrary, so that it is hard to imagine in what sense, exactly, Gyges was doing something wrong. Here is how he candidly puts it to Socrates:

> No man can be imagined to be of such an iron nature that he would stand fast in justice. No man would keep his hands off what was not his own when he could safely take what he liked out of the market, or go into houses and lie with any one at his pleasure, or kill or release from prison whom he would, and in all respects be like a god among men. . . . And this we may truly affirm to be a great proof that a man is just, not willingly or because he thinks that justice is any good to him individually, but of necessity, for wherever any one thinks that he can safely be unjust, there he is unjust. . . . If you could imagine any one obtaining this power of becoming invisible, and

never doing any wrong or touching what was another's, he would be thought by the lookers-on to be a most wretched idiot, although they would praise him to one another's faces, and keep up appearances with one another from a fear that they too might suffer injustice.

Socrates responds to Glaucon by elaborating on ideas that we encountered in Chapter 5 when we talked about virtue ethics (a common approach in ancient Greece, most famously elaborated upon by Socrates's "grand-student," Aristotle). In essence, Socrates argues, Gyges may be materially successful, but he is also morally corrupt and therefore unhappy by definition (according to the virtue ethicist's concept of happiness). Socrates saw Gyges and people like him as literally sick in their souls, incapable of truly flourishing as human beings.

By now we have learned quite a bit about the science of moral behavior, from both a neurobiological perspective (Chapter 3) and an evolutionary one (Chapter 4), so we have partial answers to Glaucon's question. As it turns out, a sense of fairness (Chapter 14) is hardwired in our brains, probably as a result of the fact that we evolved as highly intelligent social animals whose societies would simply collapse if people started behaving like Gyges and could get away with it most of the time. In an interesting sense—though surely not exactly the one he meant—Socrates was right in seeing Gyges as literally sick and therefore incapable of true happiness.

Nonetheless, both modern scientists and philosophers still struggle with the conceptual issues posed by what nowadays is called "the free-rider problem." The modern version of the problem is usually presented in less colorful (and gruesome) terms than those chosen by Plato, but the logic of it is the same nonetheless. At issue is the idea that in a society we often need to take collective action to safeguard or replenish a common resource—say, to clean up the environment, or maintain public schools, or strengthen our defense forces. This is done—ideally—by every member contributing a little (for example, through taxation) in order to reap the communal benefits. However, the mathematical theory of the so-called n-prisoner's dilemma shows that the larger the number of individuals involved (n), the more incentive there is to cheat the system and keep reaping benefits without contributing to the resource. This behavior quickly spreads, leading to the so-called tragedy of the commons: if everyone (or even just a large enough number of people) becomes a free rider, there won't be a "ride" left for anyone.

Here is how philosopher David Hume put it in his *Treatise of Human Nature* (1739):

> Two neighbours may agree to drain a meadow, which they possess in common; because 'tis easy for them to know each other's mind; and each must perceive, that the immediate consequence of his failing in his part, is, the abandoning the whole project. But 'tis very difficult, and indeed impossible, that a thousand persons shou'd agree in any such action; it being difficult for them to concert so complicated a design, and still more difficult for them to execute it; while each seeks a pretext to free himself of the trouble and expence, and would lay the whole burden on others.

Another member of the all-time who's who in philosophy, John Stuart Mill, also grasped the problem clearly when he argued (in his *Principles of Political Economy*, 1848) that the only way to reduce weekly working hours to a humanly acceptable level was to pass laws *prohibiting* people from working more than a maximum number of hours per week. Otherwise, there would be an incentive for individuals to work more than the agreed maximum, which would penalize everyone else, quickly leading to pressure on all workers to give up their right to a reasonable workweek and to labor like slaves. To this day this tension is at the root of the constant back-and-forth between labor unions and employers.

The free-rider problem also has some major consequences that even some political scientists and philosophers tend not to appreciate. For instance, it is possibly the single most powerful argument against Marxist theories of class struggle. The problem is that the better the situation becomes for the working class (because of its struggle against capitalist pressures), the better off workers find themselves. Once enough workers have crossed into middle-class status, their incentive to engage in further struggle (let alone a revolution) vanishes, and things settle into an equilibrium similar to that of many modern Western societies—which explains why repeated predictions of the coming revolution have so far abysmally failed.

Indeed, failure to appreciate the free-rider problem seems to be rooted in a common logical fallacy, the "fallacy of composition"—the assumption that the characteristics of a group must be the same as the characteristics of the individual members of that group. For instance, if there is reason to believe that cooperation is good for the group, many people infer that therefore cooperation is good for individuals within the group, but this simply does not logically follow—as demonstrated by the free-rider problem!

Despite the persistence of the free-rider issue, it is clear that societies have, by and large, been able to deal with it somehow. So we need explanations for how it is that people tend to cooperate with each other regardless of the apparently overwhelming logical force of Gyges's story. The most obvious answer is that we have governments that enforce certain types of collective cooperation, like paying taxes. Indeed, ever since Thomas Hobbes's *Leviathan* (1651), our need for collective cooperation has been presented as one main reason to support the formation and continuation of governments with significant powers of enforcement. This explanation is a crucial one, but it cannot account for all instances of willing human cooperation, since in several areas of our social experience we see people cooperate even though their efforts are not secured by the threat of government intervention. Political and social scientists have therefore considered three additional, nonmutually exclusive, explanations: flawed logic, by-product, and not just self-interest.

According to the "flawed logic" hypothesis, many people engage in cooperative activities simply because they don't understand the logic of the free-rider problem. Since there is ample empirical evidence for this lack of understanding, this is certainly a possibility, though it's hard to imagine that people never understand situations when it would be advantageous for them not to pitch in, to let someone else do the work. The "by-product" hypothesis also enjoys empirical support. The idea is that people are willing to contribute to resource X because they want something else, Y, which they can get only by contributing to X. For instance, it used to be (and in many cases still is) that labor unions provided better health care to their members than non-unionized employees were likely to get. In that case, it made sense for someone to join a union even if he despised the whole idea of unions, because of the unique benefits he would gain from joining. (Never mind the hypocrisy on which this behavior would be founded.)

Finally, the "not just self-interest" hypothesis concedes that human beings are capable of genuine acts of compassion or altruism even when they know perfectly well that they will not gain personally from such acts. Philosopher Russell Hardin cites the case of people who campaign to abolish the death penalty: their commitment to that cause can hardly be explained by their personal fear of eventually ending up on death row.

Still, all in all the surest solution to the free-rider problem is to have a system of rules in place to punish the wannabe free-riders, as in the case of a government's laws. Some fascinating empirical research has shown that people may choose such a system even when given the opportunity to opt for a different one. For instance, in a paper published in *Science* magazine, Ozgur Gurerk, Bernd Irlenbusch, and Bettina Rockenbach discuss an experiment in which they set up two fictitious institutions and gave subjects a choice of which to join. In both institutions, members would contribute to a common pool of resources, which would then be redistributed equally, regardless of the level of individual contribution. The difference was that in one institution subjects could skip the contribution without retribution, while in the second one other members would have the power to sanction the free riders, though the sanction would cost them. (In other words, the situation was analogous to paying taxes to maintain a police force and justice system.)

Not surprisingly, most subjects initially picked the nonsanctioning institution (thus demonstrating that they did appreciate the free-rider problem and took advantage of it!). The trouble was that, of course, pretty soon that institution settled at an equilibrium where the public good had been abandoned. A perfect example of the tragedy of the commons. Meanwhile, subjects who picked the sanctioning institution quickly developed a thriving system that achieved the maximum level of cooperation and resources allowed by the rules. And interestingly, more and more people switched to the sanctioning institution once they realized the advantages of its system. Hobbes would have been gratified by this mini recreation of the social contract.

So much for the science of justice and cooperation. What does modern philosophy have to say about the way things ought to be? (As opposed to either how they are or how we may go about changing them.) Throughout this book, I have argued that the relationship between science and philosophy in guiding our lives is complex, but surely one way to understand sci-phi is to let philosophy (informed by science) guide us in principle, and to use science (steered by philosophy) as our best bet for implementing those principles.

A detailed discussion of political philosophy is obviously beyond the scope of this short guide to life, the universe, and everything, but I would be remiss if I did not present in some detail what was arguably the most important contribution to the field during the twentieth century, a theory that constitutes a benchmark for any further discussion in political philosophy and unifies the current chapter and the previous one: John Rawls's idea of justice as fairness.

Rawls employs a powerful philosophical method for his analysis, one that we encountered in Chapter 14: reflective equilibrium. To review, the basic idea is that we want to strive as much as possible to harmonize our beliefs and yet recognize that they sometimes contradict each other. In practicing reflective equilibrium, we begin either with a general belief (say, a general ethical principle) we think we hold or with a specific position we have about an issue. We then ask ourselves whether the two match, and if they do not, we investigate whether it makes sense to change our position on the issue or revisit our endorsement of the general principle. This exercise can then be repeated for any set of beliefs—local as well as general—that we care about. The goal is not to achieve a (probably impossible) perfect harmony and complete logical consistency among all our positions, but rather to learn and reflect about what we believe and why, and to begin to modify some of our beliefs once we are aware of how they contradict our general view of the world.

In Chapter 14, we applied reflective equilibrium to the question of abortion. Let's now take another example dealing with someone who has strong religious-based moral beliefs. Your (hypothetical) friend Bill believes that he should obey all commandments in the Bible (both New and Old Testament). He also happens to think that while adultery is immoral, adulterers should not be killed. (Let us assume that this belief is not simply self-serving and that Bill has never betrayed his wife.) Finally, Bill discovers that a commandment in Deuteronomy 22:22 ("If a man is found sleeping with another man's wife, both the man who slept with her and the woman must die") appears to be flatly contradicted by Jesus's defense of an adulteress in John 8:7 ("He that is without sin among you, let him first cast a stone at her"). What is Bill supposed to do with this jumble of information, given that it forms an incoherent set of beliefs about adultery and the Bible?

He has three options: First, he could abandon his belief in the moral consistency of the Bible and accept that—for whatever reason—the New Testament may sometimes contradict the Old Testament. (Although then he is faced with the practical issue of which to believe, as well as with the theological conundrum of why God would contradict himself in different scriptures.) Second, Bill could accept (as some Christians do) that, because it was written later, the New Testament supersedes the Old Testament, though he would again face a significant theological problem. Finally, he could change his attitude about adultery and begin to advocate the killing of adulterers. Any way he moves within the logical space so outlined, he is using reflective equilibrium as his navigating principle. Ultimately, the outcome could be a hardening of his moral convictions, or a loss of faith in his God, or something in between. The point is that it was the exercise of trying to square his contradictory beliefs that allowed him to explicitly face them and begin to question the soundness of at least some of them. Remember, according to Socrates, the unexamined life is not worth living.

So, keep the reflective equilibrium idea in mind as we continue our discussion of what exactly Rawls is proposing. One of his starting points in *A Theory of Justice* is that a pluralist society is unable to build a system based on a single comprehensive moral doctrine. For instance, and despite much clamor to the contrary in certain conservative quarters, the United States is not—nor has it ever been—a "Christian nation." It cannot be, not only because the notion of a theocratic foundation to the US Constitution and Bill of Rights would violate both the spirit and the letter of those documents (indeed, they were inspired by secular Enlightenment doctrines, and particularly by the political philosophy of John Locke), but because it would be grossly unfair to the many non-Christians living in the country. The same reasoning could be applied, as philosophers like to say, mutatis mutandis (necessary differences being considered) if we were to think of establishing a multicultural nation on, say, Muslim principles, or Hindu principles, or atheist principles. (Contrary to popular opinion, there is a huge difference between a secular system, which is neutral toward religion, and an atheistic one, which is obviously antireligious.)

How do we proceed, then? One of Rawls's brilliant insights is that while of course members of individual religious or ideological groups should be free to follow their religion or ideology, a societal agreement can be achieved by means of overlapping consensus. One of his examples is the idea of the separation between church and state, which can be agreed to by both religionists and atheists, even though for different reasons: the religionists may not want a single state religion, or they may be wary of too much interference by the state in their freedom of worship, while atheists may dislike state support for any religious view and what they see as its pernicious effects on society.

Another cardinal principle of Rawls's system is that public discourse in an open, democratic, and multicultural society should be conducted by using public, not sectarian, language. An example discussed by philosopher Leif Wenar is that of debates about abortion. When a legislator proposes or votes on a bill concerning abortion, or when a judge rules on a particular law or case concerning the matter, according to Rawls they should do it using ideologically neutral language. For instance, it is not reasonable for a judge to write an opinion on a case invoking what God told him or justifying his position on the basis of scripture, because his opinion will apply to all members of the polis, not just to Christians, and of course some of these members will simply (and reasonably, from their point of view) reject any reasoning based on an all-encompassing moral doctrine that they happen not to share. Indeed, most judges (and to a far less extent legislators) in the United States tend to conform to this principle—witness any recent decision of the Supreme Court.

It is important to understand that Rawls is not trying to limit the free speech of any particular group, nor is he saying that individuals' moral reasoning should not be informed by their own ideologies. Of course it should. But if we take seriously the idea of a multicultural democracy, we ought to be able to *translate* our ideological thinking into neutral language that can be used by everyone as the basis for further discussion. An atheist or a member of a different religious sect would be happy to engage the Christian in debates about abortion based on neutral concepts such as the protection of innocent lives, personhood, and the like. But she would have nothing to say to someone who claims that abortion is immoral on the sole ground that (their particular) God says so.

We are finally ready to tackle Rawls's fundamental concept of justice as fairness. He acknowledges that thinking about an ideal (or simply a better) society confronts us with a constant trade-off between individual liberties and equality. Libertarians wish to maximize the former, and liberals (please notice the common root of the two words, both originating from the concept of freedom) emphasize the latter. What we want to do as a society attempting to better itself, then, is to examine the basic structure of our polis, which determines both the extent of our liberties and the degree of equality among citizens. Rawls lists a number of criteria under the general heading of "basic structure," none of which are too controversial, at least not in open Western democracies: basic rights, degree and type of opportunities, type of work and its compensation, wealth and income, access to education and health care, and the like.

For Rawls, it is reasonable to begin thinking about a fair society based on two principles. His "negative" thesis stipulates that individuals do not deserve to be male, female, Caucasian, black, rich, poor, or anything else along similar lines. They just happen to be born with a particular combination of traits, and they are lucky (or unlucky) to have (or lack) the natural endowments that come with a specific social class, gender, or ethnic group. Although our first reaction to the negative principle may be somewhat skeptical, it is hard to articulate in what sense anyone deserves to be born rich, or male, or Caucasian (or anything else), so it follows that we should agree as a society not to accord special privileges to people who happen to have been born in a certain way (or to ignore particular handicaps in others).

Rawls's "positive" thesis says that social goods should be distributed equally unless an unequal distribution benefits everyone, and particularly the least advantaged. This is arguably an even more counterintuitive idea, especially for Americans (much less so for many Europeans, I venture to guess), but it does follow from the negative thesis: if we agree that people do not deserve their good or bad luck at birth, then on what grounds might we wish to accord anything other than equal access to societal resources?

It is important to understand Rawls's partial exception to the second thesis—that inequality (within limits) may be justified if it benefits society at large, and especially if it benefits the least well off. The idea is that there may be certain activities or trades that society needs and that either require special incentives for people to engage in them or cost more (in terms of training, for instance) than other activities. In those cases, then, it would be reasonable to accord more resources to the people willing to do the hard work. In practical terms, this might justify, for example, paying police or teachers more than other professionals in virtue of the necessary incentives (a police officer risks her life to protect us) or training costs (for a teacher who needs to earn a master's or PhD degree). But it would not justify, say, disproportionate pay for an athlete, whose contribution to society is only in terms of entertainment, and whose activity clearly isn't making anyone else better off beyond the sheer value of entertainment itself.

Why does Rawls expect people to ever agree to his two theses and what may follow from them? Because, he says, human beings are endowed with a basic sense of fairness, and more specifically, we all have two "moral powers" that come to us by way of being human: we have a sense of justice, and we have a capacity for conceiving the good. Rawls does not say where he thinks these moral powers come from, but we have seen by now that something like them is indeed hardwired in our brains (Chapter 3), at least in part as a result of our evolution as social primates (Chapter 4), though surely another part comes from social, not just biological, evolution.

We have finally arrived at Rawls's crucial thought experiment, his version of a social contract, and the answer to the question of what sort of society we should agree to have, given all of these considerations. Rawls's approach here is ingenious and can be appreciated even if one does not happen to agree with his particular conclusions. He invites us to imagine that we are sitting around a table to discuss the basic structure of a new society and that we represent all the people who will be a part of that society. There is a twist to this imaginary constitutional convention: a veil of ignorance. Rawls suggests that the participants in the discussion should deliberate as if they had no information about their own or their constituents' ethnicity, gender, age, health, wealth, or any other natural endowments. What they do know is what humans generally desire (safety, food, shelter, and so on), that the society they are about to agree on has resources but that they are not unlimited, and that their society will be pluralistic (there will be people of different ethnicities, gender orientations, religions, political ideologies, and so forth). Given this unique position from behind the veil of ignorance, what sort of society would we agree to put in place?

It is crucial here to appreciate what Rawls is trying to do. He is certainly not saying that actual societies will ever be built this way, just as no *actual* social contract has ever been signed or agreed to by all members of any society. (Born in a given country, we usually have little choice but to accept whatever laws regulate its society; emigration is an option accessible only to a minority of people and at any rate gives people only limited choices of where else to go.) Rather, Rawls is challenging us to imagine the sort of rules we would be willing to build into society if we did not know in advance that we were at an advantage (or disadvantage) over others because of simple luck at birth. Remember that, for Rawls, one's natural endowments and conditions of birth aren't the sort of thing to be either morally proud of or ashamed of, because they are the result of a lottery, not of one's doing.

Now, utilitarians (reflecting the dominant position in political philosophy before Rawls's book) would argue that we should of course maximize the happiness of as many people as possible. But Rawls answers that this strategy is likely to result in unacceptable restrictions of the rights of one or more minorities. Instead, Rawls argues, the veil of ignorance will foster the adoption of a "maximin" criterion, whereby people—because they don't know whether they'll end up winning the lottery and being one of the few lucky beneficiaries of either exceptional natural endowments or birth in a privileged gender, ethnicity, or social class—will want to maximize the minimum level of resources that all have access to.

The resulting society will look neither like a welfare state (because too much control would end up in the hands of a small elite) nor like a libertarian society founded on laissez-faire capitalism (a society in which wealth and power would probably be even more skewed). It won't even be a socialist system, since too much control would be arrogated by the state. Instead, Rawls concludes, we will have either a property-owning democracy or a social democracy—in other words, a state close to the actual situation in some European (particularly Scandinavian) countries. Naturally, it is perfectly possible to object to such a conclusion. What is harder to do is to rationally justify in what sense any other society would be better than this one, as long as we agree that justice essentially means fairness.

PART VI

WHAT ABOUT GOD?

CHAPTER 16

YOUR BRAIN ON GOD

If there were no God, it would have been necessary to invent him.

—Voltaire

At the end of the 1990s, Jeff Schimmel, a Los Angeles–based writer who had a conservative Jewish upbringing, had a tumor removed from the left temporal lobe of his brain. The operation was a success in many respects, but it led to a profound alteration of his personality. Schimmel began to have the impression that people sometimes were a bit unreal, as if they were animated figures; he started hearing voices in his head, and then he had visions. He interpreted one of these visions as an appearance of the Virgin Mary, and the irony of a Jew being visited by that icon of Catholicism was not lost on him. Schimmel went back to his neurologist and underwent a new MRI scan to compare his brain before and after the operation: it was remarkably different. The affected lobe had shrunk, changed shape, and become covered with scarified tissue. That tissue was responsible for his visions and auditory hallucinations, because the scars had started causing random neuronal firings, essentially giving Schimmel a case of temporal lobe epilepsy. His brain had been turned on to religion.

Schimmel's case was certainly not the first time that a connection had been found between a misfiring brain and religious experiences. Indeed, two and a half millennia ago none other than Hippocrates labeled epilepsy "the sacred disease." Schimmel, however, looks at what happened to him not as a problem but as an opportunity. He feels that he is a better, more spiritual person and has taken up Buddhism (yes, I know, a Jewish Buddhist who sees the Virgin Mary . . .) to channel his newfound sense of religiosity. Regardless, this and other evidence from neurobiology and cognitive science raise the question of how much religion and gods are in our brain as opposed to "out there." Considering how important religion is to most people's sense of meaning and general outlook on life, we will be well served by embarking on a little investigation of the biological basis of religious belief. Who knows, we might find ways to become better human beings without having to have seizures—or even resorting to religion at all.

As it turns out, science can induce the subjective feeling of a mystical experience in a number of ways—and without having to use illegal drugs. Perhaps the most famous series of experiments in this area are those conducted by Michael Persinger at Laurentian University in Canada. Persinger has invented a device that attempts to repeat—under safe and controlled conditions—precisely the sort of thing that happened rather serendipitously to Jeff Schimmel. Persinger calls it "the God helmet": a modified motorcycle helmet capable of generating small and highly localized electromagnetic fields that stimulate specific areas of the brain within the right temporal lobe, causing a variety of responses in the subjects, including the sensing of a presence even though they are all alone in the experiment room.

If the appearance of the supernatural in your room is not your cup of tea, then neuroscientists can trigger an out-of-body experience for you, another oft-recounted and highly emotional phenomenon that many invoke as evidence of the existence of a nonphysical reality. Olaf Blanke and his collaborators at the University Hospitals of Geneva and Lausanne in Switzerland used electrodes to stimulate different parts of the brain with the goal of treating epilepsy. They found that when they induced currents in an area known as the right angular gyrus, the subject experienced what they called "whole-body displacements," or out-of-body experiences. The researchers concluded that these are caused whenever the brain fails to integrate somatosensory and vestibular information, that is, when your sense of your body's position does not square with your assessment of balance. It's interesting to note that a similar impairment of so-called proprioception (the inner sense that tells us where our body ends and the rest of the universe begins) is also caused by a variety of other stimuli often associated with mystical experiences, including abstinence from food, ingestion of hallucinatory drugs, and—most tellingly—deep prayer or meditation (hence the sense of "being at one with the universe" often reported in these cases).

Of course, the ability of science to replicate a variety of mystical experiences doesn't preclude the existence of gods and transcendental realms. After all, if we are capable of having real mystical experiences, somehow that would have to involve our brains, since that's the way we experience everything—be it a physical object or a hallucination. Indeed, more broadly, it is simply not possible to demonstrate, either scientifically or philosophically, that there is no supernatural "out there" (that is, independent of the human mind). Still, it seems reasonable to think that the more science looks into mysticism, and the more it finds ways to explain the appearance of mystical experiences, the more we are rationally compelled to lean toward the conclusion that these experiences are a result of the human brain (mal-)functioning under unusual conditions.

Here is another way to look at the problem. Suppose someone claims to have observed a flying saucer and describes its color, shape, and trajectory in detail. You do some investigating and discover that an unusually large meteor was observed at about the same time and location. Moreover, the meteor had the same color and followed the reported flight pattern. Our witness—when told about the meteor—can easily reply: "Yes, but that doesn't *prove* that there was no flying saucer. Just because you have a naturalistic explanation of what I saw, it doesn't mean that aliens do not exist." Notice what is going on here. To begin with, the reason to prefer the meteor explanation to the flying saucer one is *not* because it *proves* that there was no flying saucer. Science doesn't work that way. Rather, it is *more reasonable* to accept the naturalistic explanation because: (1) there is one explanation available that fits the facts very well, and (2) concluding that there was a real spaceship out there is an extraordinary claim and the available evidence is simply not commensurate to that claim. That said, of course it does not follow from the explanation of a particular incident that "aliens do not exist" (in general). But again, a belief in aliens (or any other belief, for that matter) becomes rational only when it is based on compelling facts. Possibilities are simply not enough.

Despite all the scientific evidence and philosophical talk about the proportionality between belief and evidence, the fact remains that a large number of people have a hard time letting go of the mystical and the supernatural. Why? This is the question we will examine in the remainder of this chapter and in the next one, seeking answers at several levels of analysis: the chemistry of the brain; the psychology and sociology of the human condition; and even the evolutionary biology of *Homo sapiens*, our own species. This is an important quest not only because our view of life and its meaning is profoundly different depending on whether or not we espouse supernatural beliefs, but because our lives are affected—whether we like it or not—by the large number of our fellow human beings who do believe in gods, or even simply in the ability of mysterious forces to alter their destiny.

Let us start with what goes on inside our heads, and in particular with the surprising effect that simple brain chemicals can have on how superstitious we are. Peter Brugger is a neurologist at the University Hospital in Zurich (Switzerland), and he carried out an intriguing experiment to see what the difference was, brain-wise, between skeptics and believers in paranormal or transcendental phenomena. It is well known that people who are inclined toward that sort of belief also have a tendency to "recognize" patterns in what are actually random bits of information. People characterized by a more skeptical mind, on the other hand, sometimes miss real patterns because they have a relatively high bar for accepting a positive conclusion. Accordingly, Brugger showed that believers often saw words or faces in what were actually meaningless patterns, and that skeptics did not recognize some words or faces when they were in fact there.

Here is the astounding part of Brugger's findings, however: he and his colleagues administered L-dopa, a substance that is normally given to Parkinson's patients because it increases the level of dopamine (a neurotransmitter) in the brain. After taking the drug, skeptics saw more faces and words than before, and their response to the experiment became much closer to that of believers! Interestingly, the drug did not further increase the tendency of believers to see patterns where there were none, perhaps because there is a plateau, an upper limit of inducible "superstition" in people. Now, why would dopamine have anything to do with our tendency to see patterns in the world around us? A clue might be found in the fact that dopamine is part of our brain's reward system: it causes a self-induced natural high when we do something right. Finding patterns to help us understand and navigate the world is generally a good thing, so our brain rewards us for it. Brugger's experiment simply shows that there is natural variation in the link between dopamine reward and the tendency to see patterns: highly superstitious people and skeptics occupy the extremes of a distribution that finds the rest of us scattered in the middle.

As it turns out, however, it's not just pleasure that is neurologically related to superstition—so is fear. Jennifer Whitson of the University of Texas at Austin and Adam Galinsky of Northwestern University elegantly demonstrated that lack of control over a given situation—a rather distressing condition for most of us—increases our tendency to be superstitious. The feeling that we cannot control what is happening to us activates an area of the brain known as the amygdala, which, as we have seen, is already closely connected with our emotional responses, and in particular with fear. However, picking up a pattern in what is going on allows us to make sense of the situation and perhaps even make predictions about future developments. This in turn diminishes our fears—as it turns out, regardless of whether our newfound "control" over the situation is based in reality or not.

There is suggestive evidence of the correlation between superstition and lack of control in the cultural anthropological literature. For instance, tribes of the Trobriand Islands (Papua New Guinea) display a conspicuously more ritualistic behavior if they tend to fish in deep rather than shallow waters. Why? The first situation is far more unpredictable than the latter (because of sudden storms, for instance, and because the fishing grounds are generally less familiar), and the increased degree of superstition apparently compensates for the increased amount of uncertainty and fear. The same effect can be seen in sports: it is a well-known fact that baseball pitchers are more superstitious than fielders, which makes sense considering that the outcome of any given play is far more predictable for the latter than the former. (Indeed, this pattern holds even for the *same player* when he switches from pitching to fielding!)

So Whitson and Galinsky tested various aspects of the connection between lack of control and the perception of illusory patterns (a form of superstition) by means of a set of six experiments in which they could manipulate both the type of pattern presented to their subjects and the feeling of control that the subjects were experiencing. In the first experiment, they established that, indeed, increased lack of control augments the subject's need to see patterns. Their second experiment went a step further by showing that this increased need to perceive patterns translates into an *actual* augmented perception of (illusory) patterns. (In this case the subjects were seeing images where in fact there was only random scatter.)

Moreover, the third experiment established that simply recalling the memory of a situation in which people experienced lack of control also increased their illusory perception. In the fourth (ingenious) experiment, Whitson and Galinsky were able to distinguish between lack of control and threat in generating superstition, and they found that threat by itself is not sufficient—it is the feeling of lacking control that generates the response. The fifth experiment focused on what happens when the lack of control is experienced in connection with financial matters, such as market volatility and investment decisions. They demonstrated that illusory correlations generated by uncertainty do drive investment decisions—clearly not a sound financial planning strategy.

The last experiment in the Whitson and Galinsky series is the one that gives us a ray of hope, as it addressed the question of how to break the connection between lack of control and superstition. All that was necessary was to give the subjects a chance to affirm themselves, that is, to be reminded that they were in fact capable of handling situations; when this happened, the connection between lack of control and superstition receded to baseline levels. The authors concluded that psychotherapy (in the broad sense of "talk therapy") may be one way for people to regain a sense of control over difficult situations, simply because it enables them to construct a narrative that makes sense of what is happening to them, thereby undercutting the need for superstition.

Then again, we can seriously ask what may seem a rather strange question for a book dedicated to the pursuit of science and reason: does superstition work? I don't mean to suggest that engaging in superstitious rituals (like basketball legend Michael Jordan's routine of always wearing his college shirt underneath his regular jersey for good luck) gives people special causal powers to alter their destiny. But maybe superstition works in a way similar to the placebo effect in medicine: within limits, if patients think they are taking a medication rather than a sugar pill, they actually feel better, and even some objective physiological measures of their health improve. (Before you think of relying too much on the placebo effect, however, be warned that the effects are short-term and do not happen with serious illness.)

It is to this possibility that Lysann Damisch, Barbara Stoberock, and Thomas Mussweiler of the University of Cologne turned, with surprising results. This was another multi-experiment study, and it's worth taking a brief look at the sequence of experiments and what Damisch and her colleagues were able to establish in their elegant piece of work. First off, they confirmed that "activating" superstitious behavior in their subjects—by having them perform a task with a ball that the experimenters told them was "lucky"—actually does improve performance. People who thought they were using a lucky ball scored much better than people who thought they were using a regular ball. Conclusion: superstition works! But inquiring minds wish to know—how does it work? Was it that people felt encouraged by the superstition "activation" and that extra degree of self-confidence made the difference? Apparently not: these researchers' second experiment was able to tease apart the effect of superstition from that of simple encouragement, and it turns out that it really was the superstition that made the difference.

The third experiment began to get at the actual psychological mechanism responsible for the improved performance of the test subjects. As it turns out, the only measurable difference was that the people primed by superstition also experienced a higher degree of "self-efficacy"—that is, they were more confident that they would accomplish what the experiments had set them up to do. But wait a minute: surely higher performance isn't simply a matter of self-confidence? If that were the case, this story would begin to sound a lot like those self-help gurus who insist that you can accomplish whatever you want if only you "believe in yourself." Sure enough, Damisch and her colleagues' fourth experiment went deeper, and what they discovered was that the people primed with superstition simply persisted at their goal for significantly longer than the control subjects did. That is why they performed better—they simply tried harder!

The moral of the story is that the causal chain looks something like this:

> You believe in superstition > You engage in superstitious behavior > This increases your level of self-confidence > This in turns causes you to persist longer at the task > You are therefore more likely to succeed, other things being equal.

No magic necessary, just some interesting—and a bit convoluted—human psychology. These results, of course, were obtained under somewhat contrived laboratory situations, but they are consistent with what we know happens "in the field," that is, in real human situations. For instance, Damisch and her colleagues report that there is a remarkable correlation between the degree of superstition of sports teams and their performance. This is also true for individual players on teams: the more superstitious the player, the better he or she performs. In field studies, of course, it is much harder to tease apart the causal factors, which is why it is the combination of field observations and laboratory studies that is beginning to tell us exactly how superstition works.

All of this notwithstanding, why would human beings go through the trouble of engaging in superstitious beliefs and behaviors to begin with? Why not cut to the chase by skipping the first two (or even three) steps in the causal scenario just spelled out and simply applying themselves longer and harder in order to succeed? Here is where things become more speculative, though certainly not less interesting if our goal is to make sense of the human condition. University of Michigan anthropologist Scott Atran suggests that superstition arises because of what he calls "the tragedy of cognition." The idea is that with consciousness comes the ability to understand one's present status, remember the past, and—crucially—project what might happen in the future. Unlike most (perhaps all) other animals, we know we will die, and our brains clearly recoil from the thought of permanent annihilation. Death is the ultimate situation in which we have no control, and we have seen that superstition is a way to alleviate the fear that is induced by lack of control. Death is the quintessential source of fear, despite much philosophizing about it, so our brains make up stories to reassure us that death is not really the end of it all.

We begin to do this at a very young age. Deborah Kelemen of the University of Arizona conducted research on how children see the world and confirmed that they have a tendency to see purpose (philosophers say "to project agency") everywhere, not only in animals (birds are there so that there is music) but in inanimate objects as well (rivers exist so that we can float boats on them). Moreover, Kelemen found, children are very resistant to alternative, nonteleological (non-purpose-driven) explanations—though perhaps not as resistant as some stubborn adults I have come across. The step from there to accepting claims about gods and other conscious forces controlling the universe is really very short.

Interestingly, neurobiological research shows that our brains make a distinction between inanimate and animate objects very early on, and that we automatically endow animate objects with volition. Paul Bloom of Yale says that even babies a few months old distinguish the two categories: if you show them an inanimate object behaving in a complex fashion as a person would do—for example, starting and stopping—the babies appear surprised, but no surprise reaction is elicited if the same behavior is engaged in by a person.

All of this, however, still does not answer the question of where superstitious belief, including beliefs in gods, actually comes from. It has been essentially universal among human societies throughout history and is still prevalent in something like 80 percent of the world population today. We need to get to the bottom of this if we want to embark on a more rational quest for the meaning of life. In the next chapter, then, we'll look at the sort of answer that may come from combining the insights of two of the most successful scientific disciplines of our days: cognitive science and evolutionary biology.

CHAPTER 17

THE EVOLUTION OF RELIGION

> The most common of all follies is to believe passionately in the palpably not true. It is the chief occupation of humankind.
>
> —H. L. Mencken

Here is an interesting experiment to consider: suppose you starve a pigeon to about 75 percent of its initial weight. Then you put it into a cage and release some food at regular intervals, making sure there is no connection between the behavior of the pigeon and the reception of food. A rather strange thing will happen: the pigeon becomes superstitious. B. F. Skinner famously did carry out such experiments, and he found that

> one bird was conditioned to turn counter-clockwise about the cage, making two or three turns between reinforcements. Another repeatedly thrust its head into one of the upper corners of the cage. A third developed a

"tossing" response, as if placing its head beneath an invisible bar and lifting it repeatedly. Two birds developed a pendulum motion of the head and body, in which the head was extended forward and swung from right to left with a sharp movement followed by a somewhat slower return. The body generally followed the movement and a few steps might be taken when it was extensive. Another bird was conditioned to make incomplete pecking or brushing movements directed toward but not touching the floor.

What was going on with Skinner's pigeons? They were trying to repeat whatever behavior they were engaged in right before they got the food, presumably in the (unconscious) hope that repeating that behavior would again make the food appear.

Philosophically speaking, the pigeons were committing a classic logical fallacy (again, unconsciously) known as *post hoc ergo propter hoc*, fancy Latin for "after this, therefore because of this." If you think it's funny when pigeons do it, just consider that this is arguably the most common reasoning mistake made by human beings. Anyone who wears his "lucky" shirt—or socks, or cap, or whatever—before an important event (an exam, a job interview, a sports event), on the grounds that at one time he wore it and the event had a favorable outcome, is behaving like Skinner's pigeons. If anything, the main difference between human beings and other animals when it comes to superstition is that other species abandon the silly behavior almost immediately once they figure out that it doesn't actually work. Humans, on the other hand, find endlessly fascinating ways to rationalize away the repeated failures of their behaviors, while constantly emphasizing the one or few times when they actually "worked."

Superstition—as we discussed in the last chapter—is ingrained in the human brain and is at the root of religious belief. With 4,200 or so cataloged religions, this is a phenomenon that concerns all human beings, religious or not, and one that we need to understand in order to intelligently navigate our way through the maze of human existence. To put it as our old acquaintance David Hume famously did (in *The Natural History of Religion*): "As every enquiry which regards religion is of the utmost importance, there are two questions in particular which challenge our attention, to wit, that concerning its foundation in reason, and that concerning its origin in human nature."

At the cost of disappointing some readers, I will ignore the first of Hume's questions, that of the rational foundations of religious belief. In my opinion, there are none, and that case has been made very eloquently in plenty of other books. The interested reader can easily find superb accounts of the standard arguments for and against the existence of gods, and several authors have restated and updated the case against gods more recently; thus, I feel under no compulsion to waste space and your time repeating it here. We are left with the second question raised by Hume: what are the origins of religious belief? This is a fascinating, though actually quite complex, question, and we need to be a bit more precise before proceeding.

First of all, let us make a distinction between superstitious beliefs in unseen causal entities (gods, ghosts, spirits, and so on) and religion as an organized set of beliefs accompanied and propagated by certain social structures. Clearly, the latter wouldn't exist without the former, but that does not mean that understanding the origin of superstition is the same as understanding the origin of religion. Just think of the obvious fact that, as we have seen, other animals display superstitious behavior, but as far as we know human beings are the only animals with religion. To make things yet more complicated, even understanding the origin of a phenomenon may tell us little about its *maintenance*—about why it is widespread. For instance, the popular social networking site Facebook started out as a way for Harvard alumni to keep in touch and network, but it is now maintained because—rather unexpectedly perhaps—hundreds of millions of people derive pleasure from connecting at one level or another, not just with friends and family but with a large number of perfect strangers.

Second, a phenomenon may have more than one cause, and these causes may work at several different levels, affording us multiple layers of understanding. This was famously pointed out by Aristotle, who distinguished four types of causes that we can illustrate with the following example. If you walk around Union Square in Manhattan, at the corner of Sixteenth Street you will see a statue of a distinguished gentleman in eighteenth-century clothes. If you were to ask how the statue happens to be there—what is the cause of the presence of that statue in the middle of Manhattan—according to Aristotle you should look for four different, and complementary, answers. First off, the statue is made of bronze, which answers the question of what sort of material the thing is made of (what Aristotle called the *material* cause). Second, you may ask what the statue represents (Aristotle's *formal* cause): it turns out to be a rendition of the French Marquis de Lafayette. Third, you may also naturally want to know who made the statue (Aristotle's *efficient* cause), the answer being the sculptor Frédéric-Auguste Bartholdi, the same guy who designed the Statue of Liberty. Lastly, you would probably ask why the statue is there to begin with (Aristotle's *final* cause), and the reason is that the

French government wished to thank the city of New York for its help to Paris during the Franco-Prussian War. (The French government picked Lafayette to be represented because he was a hero of the American Revolution.)

As you can see, even an apparently simple question like "How did that statue happen to be there?" admits of several levels of causal analysis. The important thing to notice is that these levels are not mutually exclusive: the four causes identified by Aristotle *all* contribute to the explanation for the existence of the statue, and do so in different ways. It's not that, say, the final cause (Why is the statue there?) somehow trumps or is more fundamental than the material cause (What is the statue made of?). Keep this in mind for the rest of this chapter, since confusion about different *types* of explanations abounds in discussions of superstition and religion.

When it comes to explaining the conjunction of superstition and religion, we are presented with three possibilities: cognitive science explanations, which, as we saw in the preceding chapter, tell us about how the human brain engages in superstitious thinking; and biological and cultural explanations, which aim to interpret religion as the result, respectively, of evolutionary and societal forces. Armed with our knowledge of Aristotle's awareness of multiple levels of causality, we can immediately recognize that these three explanations are not mutually exclusive accounts of religion, but rather different and reciprocally reinforcing levels of explanation. More explicitly, religion is made possible by the neurobiological characteristics of the human brain that make us prone to superstitious thinking; this is the answer to the question of how religion happens (and the subject of Aristotle's first three causes). But we equally want to know why religion happens (Aristotle's final cause). What sort of biological and cultural forces made it so that human beings not only share other animals' propensity for superstition but have pushed the notion so far as to make it into a fundamental and near-universal aspect of their lives?

Perhaps all this talk about causes and explanations sounds a bit weird when applied to religion. Indeed, it is the result of a relatively recent trend in science. For a long time scientists have stayed as far away as possible from religion, possibly because of the rather turbulent history of the relationship between the two (think Galileo, or even worse, Giordano Bruno), or perhaps because the topic was considered to be, well, a bit too sacred for an intrinsically secular enterprise like science. Yet a simple analogy will suffice to show that there is nothing strange or awkward about attempting a science-based understanding of religion. Consider language, another universal attribute of the human species, which is also, as far as we know, unique to *Homo sapiens*. If we wish to understand language, we can use the same three types of explanation we are about to discuss in the case of religion: cognitive science, biology, and culture.

There is no question that language is made possible by certain structures of the human brain. Plenty of fascinating literature in neurobiology, for instance, documents which areas of the brain are deputized to the workings of language and what happens if they are somehow injured by disease or accident. Broca's area is the region of the brain that appears to control the proper use of syntax and how we combine words into sentences, while Wernicke's area help us analyze and understand other people's sentences. Accordingly, damage to Broca's area results in ungrammatical but sensible speech, whereas damage to Wernicke's area produces sentences that are grammatically correct but meaningless. There is even a third area involved, the Sylvian fissure, which physically separates the typically human regions involved in language from the neurological structures that are present in other animals.

Notice, however, that while neurobiology gives us a satisfying explanation of how it is possible for us to have language, it provides no clue to two other important aspects of language: why we acquired the ability to begin with, and why there have been so many different languages in human societies. (About 7,000 recognized languages are spoken today in the world, and just as with religion, enormously varying numbers of people use them—from Mandarin, which is spoken by 845 million, to Ter Sami of the Urali Mountains, allegedly spoken by just two individuals. English, in case you were wondering, comes in third, at 328 million speakers).

That is where biological and cultural factors come in. Let us start with the cultural factors, since they are easier to understand as far as language is concerned. Reasonably abundant historical records enable us to trace the evolution of human languages over the past several thousand years. For instance, Danish, Icelandic, Norwegian, and Swedish all originated from Old Norse, and we can match the evolution of these languages with the migration patterns of the corresponding populations. We can also trace the evolution of individual words, as documented for English in every edition of the *Oxford English Dictionary*. The word *dictionary*, for instance, traces back to the early sixteenth century, and its root is the medieval Latin term *dictionarium*, which means a "manual of words"; that term, in turn, traces back to the Latin *dictio*, which means "word." Indeed, languages are still evolving: for instance, the word *meme* (which we will encounter again later) refers to a unit of cultural inheritance analogous to a *gene* (the unit of biological inheritance) and was coined in 1976 by Richard Dawkins, who used it in his bestselling book *The Selfish Gene*.

So now we have a good picture of both how language is possible and how it changes over time. But what explains why we have language to begin with? The answer cannot simply be "to communicate," because plenty of other species communicate without using language (which is defined not just as the ability to transmit meaningful sounds or signs but as having a grammar). It can also not be something along the lines of "how else could we carry out complex tasks, like sending humans to the moon?" (or building bridges, or using Facebook, or any other culturally complex activity). For most of human history we were not engaged in space travel (or bridge building, or virtual social networking), and so biological processes such as natural selection could not possibly have led to language for *those* reasons. Why did it happen then?

We don't really know. Several hypotheses have been proposed, of course, such as the idea that the ability for language evolved in order to better coordinate the hunting of large animals (though the paleontological evidence seems to show that humans rarely did that) or to facilitate social communication (but again, it is not clear why early humans needed such a complex tool to handle the social necessities of groups that were probably smaller than 150 people). It certainly wasn't because we would eventually get Shakespeare and the blogosphere. What we do know is that language evolved relatively late in the human lineage, surely after we evolved an erect posture. This is pretty clear from the fact that early hominids (such as the famous Lucy, a member of *Australopithecus afarensis*) were already bipedal and yet retained a small brain that probably lacked the anatomical areas associated with language. Indeed, it took us a long time to get there: the genus

Homo (to which our species belongs) began its evolution something like 2.5 million years ago, but our best estimates tell us that language was developed only during the last 30,000 to 100,000 years.

As we saw in Chapter 4, it is very difficult to test hypotheses about the evolution of human behavior, particularly behaviors that are uniquely or almost uniquely human, because there are too few comparisons that can be made with other species and the fossil record is usually not very helpful. In fact, although the fossil record tells us a pretty good story about the physical characteristics connected with walking erect first and developing language later (in terms of the development of not only human brain size but also the voice box, the anatomical feature that allows us to articulate sounds in order to speak), it has little or nothing to say about the evolutionary processes that led to the behavioral characteristic of being able to talk. As we shall see in a minute, this limitation on what we can reasonably infer about evolution applies also to the study of the evolution of religion, and for similar reasons.

Much has been written of late about the biological and cultural evolution of religion, and I certainly cannot do justice to even the popular, let alone the technical, literature on the topic here. Nonetheless, the first thing to note probably is that to talk about "biological" evolution in opposition to culture is a mistake, for two reasons. One is that cultures themselves evolve, though in a different fashion from the noncultural attributes of organisms. Moreover, cultures are just as "biological" as anything else. Humans are not the only animals to have culture, if by that we mean certain types of social behavior that are not directly traceable to an animal's genetic makeup. The distinction, therefore, should be between *genetic* evolution and cultural evolution, both of which are biological in nature, just like anything else done by (biological) organisms.

Let's talk first about genetic evolution. The idea here is that the modern version of the Darwinian theory of evolution by natural selection and other means can provide insights into why we engage in certain behaviors as human beings. As we saw in Chapter 4, I think this is essentially correct: biology can inform some debates in both philosophy (in that case, ethical philosophy) and psychology. Indeed, ever since the demise of Freudianism as an overarching theory of psychology, that discipline has been left with tons of fascinating empirical data about human behavior but not much of an underlying theory to make sense of it. What I do not believe—as we shall see in a few paragraphs—is that evolutionary biology *by itself* can provide such theory.

Whenever biologists think about the genetic evolution of any trait, behavior included, they are faced with three broad categories of explanations: the trait in question evolved by natural selection because it is adaptive (it increases the organism's fitness), or it evolved by chance processes (technically known as "random drift"), or it came to be as a by-product of other characteristics, which are in themselves adaptive. Consider some examples, such as the fact that we have a heart to pump blood; this is most certainly the result of natural selection. The heart is a complex organ with a specific and vital function, so there is no way that it could have evolved by chance, nor is it likely that it arose as a by-product of some other characteristic. Genetic drift, on the other hand, is probably responsible for the type of variation that simply does not matter to survival: for instance, the number of hairs on men's chests. It may be important, in certain climates, to be particularly hairy, but the precise number of hairs probably makes no difference, and so the trait evolves by random sorting of whatever genes cause different degrees of hairiness. What about by-products? A good example is represented again by the heart, this time not by its main function but by the fact that it makes a noise when it pumps. The noise is not necessary to survival and was therefore not selected for by natural selection. It is there because pumps make noise and there is no way to avoid that; it is therefore a by-product of the evolution of something else (the ability to pump blood).

What about religion? We can safely exclude drift as a possible cause. The notion is well established in biology that drift does not lead to the evolution of complex and "expensive" (in terms of energy) structures, and religion unquestionably fits that bill. It must be either selection or by-product. In terms of selection, several authors have proposed that religion might have been favored because it fosters prosocial behavior. The idea is that groups of individuals that are more cohesive, whose members are more willing to help and sacrifice for other members of the group, outcompete groups whose members are less inclined toward cohesive behavior. In a sense, religion is a type of social glue that makes people more collaborative by way of a combination of threats and rewards (such as eternal damnation or a permanently blissful afterlife). It is very hard to evaluate this sort of hypothesis empirically. For instance, the evidence that religious communes in nineteenth-century America persisted longer than secular communes would seem to support this hypothesis. Then again, once researchers statistically controlled for different degrees of costly requirements for membership in a commune, the effect disappeared—suggesting that it is having to pay a high price to belong to a group that makes people stick with it, not religiosity per se. Moreover, the "prosocial behavior" hypothesis relies on a form of group selection (natural selection favors *groups* with certain characteristics, not individuals), and there are good technical reasons in evolutionary theory for being skeptical that group selection actually happens very frequently in nature. Finally, it should be pointed out that other primates also exhibit prosocial behaviors, which are enforced through a set of rewards and punishments meted out by other members of the group, presumably without the necessity to develop religions.

According to another possible scenario, natural selection could have favored the evolution of religion through the standard mechanism of selection to augment individual (as opposed to group) fitness. As we saw in the last chapter, engaging in superstitious behavior tends to alleviate anxiety and stress because it makes one feel somewhat in control (however illusory such control may be) of whatever is causing the anxiety or stress. As we also discussed, arguably the most stressful thought we have throughout our lives is that of our own demise. We referred to it as the "tragedy of cognition": once an animal is conscious of its existence as independent from the rest of the universe, it also becomes conscious of the fact that such existence is limited—and few people can contemplate permanent personal annihilation with the equanimity of Epicurus, who famously said, "Death does not concern us, because as long as we exist, death is not here. And when it does come, we no longer exist." That religion evolved to soothe our brain when it contemplates its own annihilation is therefore an appealing possibility, backed up by some degree of neurobiological evidence. But there are several objections that can be raised against it, the most obvious of which is that perhaps the evolution of *superstition* can be explained that way, but religion is a much more complex social—not just individual—phenomenon; while rooted in superstition (in the sense that religion would not be possible without it), religion goes well beyond it. Natural selection is a rather minimalist sort of process: it produces results that are good enough for what needs to be done, so it is hard to imagine that the highly complex social phenomenon of religion would evolve if the simple individual phenomenon of superstition could suffice.

Of course, there are many other possible scenarios linking natural selection and religion as an adaptive trait, but we've got enough of a flavor of that possibility to turn now to its most convincing potential rival explanation: the idea that religion did evolve, but as a by-product of something else. Of what, exactly? There are two characteristics of the human mind that are very good candidates to make us prone to superstitious belief and may have started *Homo sapiens* down the road that led to the cultural phenomenon of religion. The first such characteristic, our ability to pick up nonrandom patterns in the world around us, is shared by a large number of animal species. That ability is obviously advantageous for the survival of an animal. It is necessary, for instance, to figure out the succession of seasons to know what to hunt or gather at different times of the year (and later in human evolution, when to sow seeds and harvest crops). It is clearly good to be able to tell which places are more likely to provide food or water and where potential predators may lurk.

However, there is a major problem with pattern seeking, one that we saw in the last chapter when we were talking about how high levels of dopamine increase our tendency to see patterns where there are none: inevitably an animal will make a mistake and confuse random noise for meaningful information. Natural selection hasn't eliminated this type of error (which in scientific parlance is called a "false positive") because it probably isn't very costly, especially when compared to its opposite—ignoring a signal thinking that it is noise (called a "false negative"). The typical example brought up in the context of this discussion is that if you hear a rattling of leaves nearby, you can judge it to be the result of either wind (random signal) or a predator lurking nearby (meaningful pattern). If you go for predator and it is in fact the wind, not much harm is done, other than getting a scare and wasting a bit of energy engaged in an escape maneuver. But if you think it is the wind when in fact there's a predator ready to attack, that judgment could literally be the last mistake you ever make in your life. (Incidentally, one naturally wonders why there has to be a trade-off between false positives and false negatives: isn't there a way to minimize the likelihood of both? As it turns out, there is, but it requires gathering additional information, which in itself is a costly and potentially risky enterprise.) We have already seen that pattern-seeking behavior generates superstition-like conditions in other animals, as shown by several studies demonstrating how widespread it is in our species. You can think of this first pillar of superstition, then, as the result of an imperfect mechanism for attributing causality: sometimes we imagine specific causes for things that are actually the result of random or not particularly meaningful processes.

The second characteristic that may have made us prone to superstition and eventually religion is the fact that human beings (and perhaps a few other primates) have what is called a "theory of mind" engrained in their behavioral repertoire. This is not really a "theory" in the scientific sense of the word, but the very useful ability (for social beings) to project agency onto others in order to understand, predict, and correctly react to their actions. Just as with pattern-seeking behavior, agency projection can go beyond what's useful—as when we get rather irrationally mad at our computer (note, not at the people who built it or programmed it, but *at the machine*) for not responding properly to our commands.

How do we combine pattern seeking and agency projection to explain religion? The most common and simple type of religious belief is animism, the idea that natural phenomena have a soul of some sort, a type of diffuse natural divinity. It is from animism that more sophisticated conceptions of religion eventually arose, from pantheism (the identification of individual gods with natural objects, such as the moon, the sun, the planets, and other natural phenomena) all the way to polytheism and monotheism (the personification of gods in humanlike form). It is easy to see that animism may have arisen naturally from pattern seeking and agency projection, and that both these characteristics still underlie all forms of religious belief. It is in this sense that religion can be seen as a by-product of evolution, since the two behavioral traits underlying it do have adaptive significance (in the sense of augmenting an individual's fitness) and were likely to have been the result of natural selection (though this is by far more clear in the widespread case of pattern seeking—which we share with many species—than with agency projection, since we don't know much about the evolution of consciousness in prehuman species).

All of this being said, we have limited our analysis to the issue of genetic evolution of superstition and religion, be it as a direct result of natural selection or, as I think more likely, as a by-product of preexisting human characteristics. Clearly, however, religion has also been affected by a dramatic degree of cultural evolution, to which we now turn in order to complete our picture of how religion came to be such a dominant component of most people's lives. To begin with, let us establish why genetic evolution is not sufficient to explain the totality of the religious phenomenon. We can understand this through two types of consideration. First, it is easy to show that there simply isn't enough genetic variation in the human population to explain the variety of religious beliefs and practices. Perhaps this is a bit like shooting fish in a barrel, since very few people would seriously maintain that every aspect of human behavior can be accounted for entirely by our genes (though you'd be surprised!). Still, the point is that the number of gene variants affecting behavior that are present in the human population is probably much smaller than what it would take if genes really were the only, or even chiefly, causal factor as far as our beliefs are concerned, particularly our beliefs in the supernatural. There just aren't enough genes in the human genome to go around.

To better appreciate this first argument, it is helpful to introduce my second one, since the two are related. Recall my early example of the distinction between the ability to have language and the variety of actual human languages. I said that the ability to have language is genetically engrained, while the variety of languages is the result of cultural evolution. That is, our genes teach us (somehow, the details are far from being clear, despite some spectacular recent advances) how to speak, but not which language to speak. Indeed, there wouldn't be enough genes in the human genome to codify all the words in a typical advanced human language, like English, Chinese, or Spanish—let alone to genetically encode all seven thousand currently used languages (and the many more that are no longer spoken). My argument is that the difference between language ability and spoken languages nicely parallels the difference between a propensity for superstition and the bewildering cultural phenomenon of religion. If you are still not convinced, here is yet another example that makes clear the limits of genetic explanations of human cultural phenomena, of whatever nature. It is undoubtedly true that we have a natural craving for fatty and sugary foods, which in turn is likely to be genetically encoded in us so that we take advantage of every opportunity we encounter to store fats and sugars—after all, it used to make the difference between death and survival. This explanation, however, tells us next to nothing (or if you prefer, tells us only "trivial truths") about a huge range of cultural phenomena related to food, from fast food to gourmet restaurants, from cookbooks to *Top Chef*, and so on.

At this point it is pretty much de rigueur to examine the concept summarized by a word we briefly encountered at the beginning of this chapter: *meme*. A meme is supposed to be a unit of cultural evolution, and memetics is the (alleged) science that studies the evolution of memes. As we have seen, the concept (and term) can be traced back to 1976 and the publication of Dawkins's *The Selfish Gene*, which was aimed at explaining some counterintuitive aspects of evolutionary theory to the general public. Dawkins drew an analogy between memes and genes, the units of biological hereditary information, though to his credit he did not push the idea beyond the status of mere metaphor (others did so afterward).

Let me be blunt about this: I do not think memetics is in the least helpful in understanding how cultural evolution takes place. Although a thorough discussion of memetics would be beyond both the scope and the point of this book, we have to tackle the issue because you as a reader are bound to encounter it whenever the broader topic of genetics versus culture comes up. There is no question at all that cultural traits evolve, in a sense, above and beyond genetic evolution. It is also equally clear that they do so by following non-Darwinian dynamics. While genes are inherited vertically (for the most part, there being exceptions), that is, from one generation to the next, cultural traits are inherited both vertically (from parents to offspring—by far the best statistical predictor of your religious or political affiliation is your parents' religion or political sympathy) and horizontally, that is, through learning from other individuals. But the disanalogy between genes and memes runs much deeper than this.

To begin with, memes (unlike genes) are so ill defined that everything from a catchy tune that gets stuck in your head to "religion" can be called a meme, making it an extremely vague and highly heterogeneous category of cultural phenomena. Second, while we know what genes are made of (a couple of different types of nucleic acids), memes have no definable physical basis. They are ideas, after all, and ideas can be instantiated as neural patterns in someone's head, as a book, as bits stored on a computer's hard drive, or as something else entirely. Third, while we know what it means, chemically, for a gene to mutate, the analogous process for memes is once again hopelessly vague—and without an understanding of memetic mutation no scientific theory of memes can be generated. Fourth, and most crucially, the fundamental attraction of the analogy between genes and memes is that we gain an explanation (not just a description) of what goes on in cultural evolution. Except that we actually don't.

You see, what allows the theory of evolution to work as a scientific theory—as opposed to an empty truism—is that we know enough about what genes do to be able to make predictions about which genes should be favored by natural selection and which ones shouldn't. We can then go out there in the real world, test those predictions, and modify our understanding accordingly. That is the way science works.

But we have no idea at all how the "functional ecology" (to use the proper term from the biological sciences) of memes is supposed to work. We don't know why a particular tune, or a particular religion, should be favored by memetic selection while other tunes or religions shouldn't be. As a consequence, to say that a particular meme (say, the tune of the *A-Team* TV series) spreads because it is favored by selection really amounts to saying that the tune spreads because it spreads, thereby making memetic theory tautological. It is a pretty metaphor entirely devoid of actual science, though I cannot say whether this is because memetics is fundamentally flawed as an idea for a research program, or simply because it is a relatively young enterprise (thirty-five years and counting). We need to keep in mind, however, that the science of genetics at the same age had made spectacularly more progress than memetics has so far. Be that as it may, memetics is currently not helpful at all in understanding cultural evolution, and in fact it muddies the water significantly.

The general picture of the evolution of religion that I see emerging so far, then, is similar to the one we have seen for the evolution of morality, or the one that was probably responsible for the evolution of language. Genetic evolution provided the building blocks of pattern-seeking and agency projection behaviors, getting the process started as a by-product of behaviors that were adaptive for other reasons (understanding threats and opportunities in nature and being able to understand and predict other people's behavior, respectively). It is also possible—though I am not convinced by the arguments and evidence adduced thus far—that natural selection directly favored religious behavior either because it reduces stress in individual humans or because of the prosocial behavior that makes some groups more competitive than others. The move from superstitious and simple religious beliefs to the bewildering variety and complexity of modern religious cultures, however, was a result of cultural evolution, a process that takes place on top of and by distinct mechanisms from the standard genetic-Darwinian one. All of this provides a reasonable explanation for one of the most important and puzzling human phenomena, and certainly one that should seriously be considered if our goal is to form as rational an understanding of the world as is humanly possible. Needless to say, gods are not actually excluded from the picture I have painted, but they are also very clearly not required. In the next chapter, we will see that even if gods did exist, we still wouldn't need them for much that's of any importance to living a meaningful and moral life.

CHAPTER 18

EUTHYPHRO'S DILEMMA: MORALITY AS A HUMAN PROBLEM

> And what is piety, and what is impiety?
>
> —SOCRATES

I'S THE YEAR 399 BCE OR THEREABOUTS. WE ARE IN ATHENS, walking beside one of the greatest figures of Western civilization, the philosopher Socrates. As it happens, Socrates is on his way to the Agora, the main gathering place for citizens in ancient Athens. He is not going there for commerce, nor to engage in a discussion with one of his pupils. Rather, Socrates has been summoned on urgent business at the Royal Stoa, the office of King Archon (the legal magistrate). The reason for the summons is that a young Athenian named Meletus, whom Socrates hardly knows, has charged the philosopher with impiety (disrespect for the gods and general immorality) and of corrupting the Athenian youth. As we learn from Plato's *Phaedo*, Socrates's defense (described in another Platonic writing, the *Apology*) will fail, and he will be put to death by the Athenian democracy.

But that nefarious day in the history of philosophy is still ahead of us; at the moment, Socrates has encountered an acquaintance, also on his way to the magistrate's office. The character in question is Euthyphro, which is also the name of a dialogue in which Plato (who was Socrates's student and Aristotle's teacher) describes one of the most powerful arguments ever deployed to show that even if gods existed, and contrary to popular perception, they would have no role in how we decide what is moral and what is not. This is a crucial issue, because regardless of the arguments we have made so far about religion as a human rather than supernatural phenomenon, for most people a main reason for believing in God (or gods) is their feeling that only the supernatural could possibly guarantee the existence of a universal morality, and by implication that only the existence of that sort of moral code provides ultimate meaning to our existence. But if Socrates (or Plato, whose contributions are hard to distinguish from those of Socrates because the latter never wrote anything down) is right, then the question of the existence of gods is irrelevant to both morality and the quest for meaning in life—which implies that no shortcut based on sacred books will do and we need to do the sort of hard work in which we have been engaged so far.

So let's follow Socrates for a bit longer and see what happens when he encounters Euthyphro. After exchanging greetings as customary, they inquire into each other's business at the King's Court. Euthyphro is aghast that someone would file suit against Socrates, but it is Socrates who is more surprised when he finds out Euthyphro's business: the guy is going to denounce his own father, who accidentally caused the death of a household employee, who had in turn been guilty of murder. Socrates wants to know how Euthyphro can be so certain, judging from his boundless self-confidence, that this is the right course of action for him to take. Euthyphro's response is that he knows what he is about to do is right because that's what the gods want. But how, replies Socrates, do you know what the gods want? Completely unperturbed by the obvious irony in Socrates's question, his interlocutor candidly responds: "The best of Euthyphro, and that which distinguishes him, Socrates, from other men, is his exact knowledge of all such matters. What should I be good for without it?"

Socrates feigns then much reverence for Euthyphro and declares himself to be the latter's disciple, so that he too can learn about such important matters. This setup immediately leads the philosopher to ask the obvious question: "And what is piety, and what is impiety?" Remember, in modern parlance this question is about the same as asking what is moral and what is immoral. Euthyphro's first answer is one that most people would give: "Piety, then, is that which is dear to the gods, and impiety is that which is not dear to them." In other words, gods *define* what is moral or immoral. This same sort of answer is why so many people are absolutely convinced that morality cannot possibly exist without gods, and that therefore denying the supernatural is equivalent to embracing moral relativism, and from there the distance is short to the conclusion that life is meaningless.

But not so fast, says Socrates. He points out to his companion that, according to the stories we hear, the gods often disagree vehemently on what is right or wrong in any particular instance. This, of course, is a problem not just for polytheistic religions but also for monotheistic ones once we realize that the intelligent person ought to ask herself why she should embrace the moral dictates of one particular god rather than those espoused by another god of a competing religion. But Socrates this day is in a good mood, so he lets Euthyphro off that particular hook by postulating that there probably are at least some moral dictates on which all gods would agree (for example, that killing without reason is not permissible). Still, Socrates presses the point by rephrasing the question: "The point which I should first wish to understand is whether the pious or holy is beloved by the gods because it is holy, or holy because it is beloved of the gods." Let us examine these two alternatives—the horns of what is now known as "Euthyphro's dilemma"—very carefully. If you understand why the dilemma is so powerful, you will have liberated yourself from the misguided notion so common among humanity that morality and divinity are inextricably entwined.

Consider first the second horn, that something is moral because it is approved by the gods. Rather counterintuitively, this essentially means that morality is arbitrary! If God decides that, say, murder, rape, or genocide are okay, then we would have to assent, regardless of how repugnant such a thought might be or how much our own sense of right and wrong would be offended or crushed by it. Indeed, it is not at all difficult to find perfectly good examples of God's commandments in various sacred scriptures that no person in his right (moral) mind today would follow, regardless of their alleged divine origin. Let us take a few examples from the Old Testament, the sacred book shared by the Judeo-Christian-Muslim trio of Abrahamic monotheistic religions, which together account for about 55 percent of believers worldwide (followed by about 15 percent who are nonreligious or atheist and 13 percent who are Hindu):

- In Genesis 6:11–17 and 7:11–24, God exterminates nothing less than the entire human race with the exception of a single family. He also kills every other species on earth, save for a pair of individuals to repopulate the planet later on.
- In Genesis 34:13–29, the Israelites, with God's approval, kill Hamor and his son, together with the entire male population of his village, while not neglecting of course to take for themselves women, children, cattle, and other possessions.
- In Exodus 14, 9:14–16, 10:1–2, and 11:7, God brings plagues upon the Egyptians (even though there is no historical record that the Jews were ever held in captivity in Egypt). The reasons given, which do not seem particularly convincing from a moral perspective, include showing that He is the Lord and there is nobody else like him, displaying His power, and giving the Israelites something to pass on to their children.
- In Exodus 17:13, Joshua slaughters Amalek and his people, with God's approval.
- In Numbers 15:32–36, someone breaks the Sabbath by gathering sticks for a fire. One might think this a minor offense, but the God-sanctioned punishment is stoning to death.
- In Deuteronomy 2:33–34, the Israelites wipe out the men, women, and children of Sihon (needless to say, with God's approval).
- In various passages in Joshua (6:21–27, 8:22–25, 10:10–27, 10:28, 10:30, 10:32–33, 10:34–35, 10:36–37, 10:38–39), said character kills far and wide, including,

respectively, the people of Jericho and Ai, the Gibeonites, the people of Makkedah, the Libnahites, the people of Lachish, the Eglonites, the Hebronites, and the Debirites.

We could go on and on, but I think that my point (and Socrates's) has been abundantly made. Perhaps, then, we should embrace the other horn of Euthyphro's dilemma and agree that a given action is approved by the gods because it is moral, not the other way around.

Except that such an agreement provides only temporary relief. Think of it this way: if God approves of a given action because that action is moral, this means that there is a God-independent standard for morality by which God himself abides. But if that is the case, two astounding conclusions follow: first, we do not need gods to be moral; and second, we now need to figure out where morality comes from. We have already seen the answer to the latter question (in Chapters 3 and 4), but for the present purpose the surprising outcome of Euthyphro's dilemma is that the religious believer has to agree that either morality is arbitrary or the divine, even if it exists, has nothing to do with it at all.

Of course, few people like this conclusion the first time they hear it, least of all our good old friend Euthyphro, who tries desperately to escape the horns of the dilemma on which Socrates has managed to impale him. He does not succeed, and his attempts reveal such a poor logic that Socrates comments:

> And when you say this, can you wonder at your words not standing firm, but walking away? Will you accuse me of being the Daedalus who makes them walk away, not perceiving that there is another and far greater artist than

Daedalus who makes them go round in a circle, and he is yourself; for the argument, as you will perceive, comes round to the same point. Were we not saying that the holy or pious was not the same with that which is loved of the gods? Have you forgotten?

Socrates, the infinitely patient teacher (or, depending on how you interpret his character, the always sarcastic commentator on society), then tells Euthyphro that they now have to begin the discussion from scratch. But Euthyphro cannot take it anymore, and in one of the most unceremonious hasty retreats ever to appear in Western literature he takes leave of the philosopher by saying, "Another time, Socrates; for I am in a hurry, and must go now."

The *Euthyphro* dialogue was written twenty-four centuries ago, and its conclusion is devastating for the whole Judeo-Christian-Muslim conception of the relation between divinity and morality. Indeed, it is devastating for any religion that attempts to connect gods and ethics (which is pretty much all of them). One would expect, then, some energetic responses to Plato's point. Indeed, several theologians have taken up the challenge, even though my sense is that the majority of philosophers pretty much concede that Plato's position is unassailable. Since this is a crucial problem, let us take a brief look at the three standard objections to the dilemma.

Perhaps the most obvious line of counterattack was pursued first by one of the most influential theologians of all times, Thomas Aquinas (1225–1274). Aquinas accused Socrates (or Plato) of engaging in a logical fallacy (which is really bad for a philosopher), and specifically the fallacy of the false dilemma. This occurs when someone presents only two choices in a situation where in fact more options are available.

Politicians are particularly skilled at this sort of thing (as in "you are either with us or against us"). Aquinas conceded that something is good because God says so, but—the theologian went on to argue—this is simply because it is in God's nature to be good, which guarantees that his commands will in fact be moral. (One would have to conclude that Aquinas was not too familiar with the Old Testament passages just cited, or that he did not take them seriously enough.) However, this is not at all a satisfactory move, theologically speaking, because it amounts to a rejection of the divine command theory of morality that is so fundamental to religion. Moreover, it essentially impales one on the second horn of the dilemma: after all, if God *has* to act in a certain way by his own nature, then in a very real sense morality is again independent of God himself, at once demonstrating that God has limits and opening up the possibility for humans to figure out what is right and what is wrong on their own.

A somewhat more sophisticated attempt has been made by a modern-day theologian, Richard Swinburne. It takes the form of a compromise, suggesting that moral values come in two flavors: necessary and contingent. In other words, some moral rules are universal and absolute, while others depend on circumstances. Absolute values, according to Swinburne, hold in all conceivable worlds, examples being the prohibitions against rape or murder. Contingent values, on the other hand, are not applicable everywhere and at every time—let's say the prohibition on eating certain kinds of foods at particular times of the year. Swinburne's stratagem, however, hardly makes things better for the religionist: if absolute values are independent of specific circumstances, then they can be arrived at by reason (which is of course the project of most ethical philosophers), and one falls yet again on the horn of the dilemma that says we don't need gods to tell us what to do. In this scenario, God at best gets to tell us his personal preferences in terms of minor actions that, frankly, hardly seem a matter of morality at all. (I mean, who is being harmed by the fact that I decide to eat meat on Friday or work on the sabbath? Apparently, only God's vanity—and the animal from which my steak was obtained.)

Finally we come to arguably the most sophisticated response to date to Euthyphro's dilemma, the one proposed by a contemporary philosopher of religion, Robert Merrihew Adams. Adams distinguishes two meanings of words like *right* and *wrong*: one refers to what we all mean by those terms, an understanding that even an atheist can share. The second meaning is specifically religious and indicates simply what God wants, regardless of human judgment of the morality of such wants. The crucial move, then, according to Adams, is that God is by nature good (how we know this, given the evidence in certain sacred texts, is rather obscure, but let us go on for the sake of argument), which is why the two meanings of *right* (or *wrong*) actually coincide. But, according again to Adams, God could decide (indeed, in my opinion, *has* decided over and over again) to command differently, thereby separating the two meanings of *right* by making, for instance, rape, murder, and pillaging "moral" in the second sense. I don't know about you, but this sounds to me like an incredible exercise in mental gymnastics aimed at desperately avoiding the conclusion that Plato was right. Indeed, in a very circuitous way, we are back to one of the horns of the dilemma—the admission that morality is arbitrarily defined by God and that therefore anything he says must stand simply because he is so powerful that it would be foolish to resist him. But by that token, all sorts of atrocities committed by force during human history would also have to be considered in some sense "moral" because they were the result of a decision made by a very powerful individual. If that's your idea of morality, I think we've got a problem.

If we combine what science tells us about our beliefs in gods (Chapters 16 and 17) with the devastating force of Euthyphro's dilemma, we have to conclude that religion is a human phenomenon, not the reflection of a supernatural reality, and that when it comes to morality (and therefore to a big chunk of what gives meaning to our life), we are on our own. The course of action suggested by this conclusion is fairly clear: to embark on a quest to figure out what morality is and what sort of scientific and philosophical insights we can appeal to in constructing it. Which just happens to be precisely what we have been doing all along in this book! It is now time, therefore, to draw some general lessons from our sci-phi-guided exploration.

CONCLUSION

HUMAN NATURE AND THE MEANING OF LIFE

> Man is a goal-seeking animal. His life only has meaning if he is reaching out and striving for his goals.
>
> —Aristotle

We have come to the end of our tour of the science and philosophy of living meaningfully. During the journey we learned about the neurobiology of moral decision-making as well as about John Rawls's theory of justice, we have examined the hormones that make us fall in love, and we have discussed the ethics of friendship. What we have not examined yet is an assumption underlying this entire book, the same assumption that Aristotle made almost two and a half millennia ago: that there is something fundamental about being human, that we all seek pretty much the same things, though we may arrive (or fail to arrive) at them in a bewildering variety of ways. In other words, that there is such a thing as human nature.

These days the concept of human nature is as unpopular with philosophers as it is intriguing to scientists, and it will be instructive to take a look at both misguided and interesting ways of thinking about it. Perhaps the crassest science-inspired approach to human nature is the one proposed by evolutionary psychologists. Evolutionary psychology, an offshoot of evolutionary biology, is based on the very reasonable assumption that some human behavioral traits have at least a partial genetic basis and have been shaped, in part, by natural selection. As such, the claim is rather uncontroversial, but the devil is in the details. Some evolutionary psychologists are prone to make spectacularly broad claims about human nature that are almost entirely unsubstantiated by the evidence. Just to give you a flavor of what I'm talking about, here is a sampler from Alan S. Miller and Satoshi Kanazawa, whose article in *Psychology Today* is provocatively entitled "Ten Politically Incorrect Truths About Human Nature":

> Men have built (and destroyed) civilization in order to impress women, so that they might say yes.

> Thus, men who prefer to mate with blond women are unconsciously attempting to mate with younger (and hence, on average, healthier and more fecund) women.

> Muslim suicide bombing may have nothing to do with Islam or the Koran. . . . As with everything else from this perspective, it may have a lot to do with sex, or, in this case, the absence of sex.

The first statement is ridiculous on the face of it. Every other animal species on the planet has managed to get their females interested in sex without having to design rockets to get to the moon, so it isn't clear why we would be the exception (or on what bases Miller and Kanazawa have arrived at this bizarre conclusion). The second statement neglects a few obvious facts about human biological and cultural variation—for instance, that many cultures at most points in time have simply not experienced the pleasure of admiring blond women (so this clearly can't be a human universal), nor that there is any sensible reason to think that being blond is a proxy for being young and fecund. (The more likely biological explanation for why "gentlemen prefer blondes"—if any is actually needed—is that a number of animal species tend to prefer unusual-looking mates, as long as they are healthy, since this increases the chances of a higher genetic diversity in their offspring.) As for the prevalence of suicide bombers among Muslims, which Miller and Kanazawa attribute to tolerance for the practice of polygyny (which leaves some men without access to a mate), they completely discard two obvious facts: First—as they acknowledge apparently without taking in the consequences—there are plenty of other polygynous cultures in the world that don't engage in suicide bombings. Second, suicide bombing is a very recent cultural development, unknown until a few decades ago, and therefore hard to trace to any deeply engrained biological tendency.

So much for evolutionary psychology, but are there more sensible ways of thinking about human nature that are informed by reasonable philosophical insights and reliable scientific evidence? Indeed there are. Philosophically speaking, an interesting take was proposed by David Hume in his aptly titled *Treatise of Human Nature*, published in 1739–1740. Hume got involved in a dispute that was going on at the time between two different schools of "originalism," the idea that human nature is what it is and does not change over time. One school, represented by Francis Hutcheson and Anthony Ashley Cooper, maintained that human beings are naturally benevolent and that this accounts for our sociability. The other school, championed by Bernard Mandeville, defended the position that human sociability arises out of self-interest. Hume, very sensibly, struck a middle ground, but in an ingenious way that foreshadowed the recent findings of the best twenty-first century science.

As pointed out by Michael Gill in his insightful analysis of the Scottish philosopher's writings, Hume's idea was that human nature has an "original root" (today we would say a "biological basis") that is in fact largely concerned with self-interest and that provided for a minimal sociability early on in our history. But "society" (we would say "culture") builds on this foundation, expanding the reach of our genuine feelings of concern from our immediate kin and neighbors to the ampler and more abstract circle of humanity at large. As Hume puts it: "Thus self-interest is the original motive to the establishment of justice; but a sympathy with public interest is the source of the moral approbation which attends that virtue." This amounts to a dynamic (and intrinsically hopeful!) view of human nature, one that in a sense echoes Aristotle's virtue ethics (Chapter 5) and his contention that virtue is a matter of practice. If that were not the case, everything you've read so far in this book would be of academic interest, but it wouldn't help you in pursuing a eudaimonic life.

What about science? In 1953 James Watson and Francis Crick published a paper in *Nature* magazine entitled "A Structure for Deoxyribose Nucleic Acid." It detailed their discovery of the structure of DNA, the hereditary molecule used by most living organisms on our planet (some use a different but similar one, called RNA). It was also the beginning of the so-called molecular revolution, a period of intense discoveries about the chemical basis of life. The revolution arguably culminated in 2003, with the publication of a complete draft of the human genome project. The project, which had started in 1989 and had cost about $3 billion, was heralded as the point at which science would finally uncover human nature, when we would be able to literally read on a CD how to make a human being—which of course immediately carried the prospect of curing all sorts of diseases, from cancer to aging itself.

By the end of the first decade of the twenty-first century, most scientists had become a bit more sober about this, humbled not just by the enormous complexity of the human genome itself but by the increasing realization of what should have been obvious from the beginning: it's not all in our genes, far from it. Don't get me wrong: biology, and genetics in particular, is fundamental to understanding who we are. We have large brains because of genes that make 'em so, and that in turn is the result of evolutionary processes (though, remarkably, we still have pretty fuzzy ideas about exactly why evolution favored such a ridiculously expensive device, metabolically speaking, as the human brain). But it is increasingly clear that what has mattered most for human evolution and the shaping of human nature during the past tens of thousands of years wasn't genes, but culture (as well as how the two interact, the so-called gene-culture coevolution).

Still, once both the human and chimp genome projects were completed, biologists thought we were very close to getting our hands on the biological basis of human nature: just compare the two sets of genomes and identify the differences. Since chimps are our closest evolutionary cousins still alive today (though separated from us by a whopping four million years of evolution), we should be able to pinpoint the genes that make us human. Again, not so fast, as it turns out. Researchers have found a surprisingly small number of proteins that appear to have undergone rapid evolution since the separation from our chimplike ancestors. One of the exceptions is a protein called FOXP2, which is involved in human speech. Some differences in regulatory sequences, like a small RNA molecule called HAR1, have also been found, though all we know about it is that it is expressed in fetal brain cells—we have no idea what it actually does.

More remarkably, the clearest burst of genetic change marking human evolution happened much, much more recently than the time of our splitting from the lineage that led to chimpanzees: between only 30,000 and 100,000 years ago. That's puny in evolutionary terms, and clearly within a period when cultural evolution was highly relevant. (Thirty thousand years ago corresponds roughly with the evolution of language, and 100,000 years ago with the invention of agriculture.) This is a major reason why an increasing number of scientists think that human nature has been shaped by culture, acting via a feedback loop on our genetics. The obvious example is the repeated evolution of genes that produce proteins allowing human beings to metabolize lactose, which is found in milk and sugar. Not at all coincidentally, these mutations were favored in populations that had started to raise cattle—a crystal-clear example of culture guiding genetics. Hume would have been pleased.

Indeed, much of what we have learned about what makes our lives meaningful and ourselves happy does not come from molecular or evolutionary biology, but from the social sciences, particularly psychology and sociology. Modern psychology makes the same distinction that Aristotle made between happiness as the simple pursuit of pleasure and happiness as a fulsome, eudaimonic life. The empirical results are clear and would not have surprised the Greek philosopher a bit: seeking pleasure for pleasure's sake (what psychologists call the "hedonic treadmill") doesn't lead us anywhere because as soon as one pleasure is achieved another one appears over the horizon. The quest never ends and—more importantly—never truly satisfies. Of course, that hasn't stopped modern American society from essentially turning into a gigantic hedonic treadmill powered by advertising and fueling corporate profits. But perhaps that is one major reason why so many people in this country find life unsatisfactory and turn to mood-altering drugs (which are heavily advertised and produce large corporate profits). Psychologists have found that what really satisfies people instead is lifelong happiness—which comes only through the search for meaning.

Modern science has also found that our emotions can be worked out as if they were muscles, in a fashion not very dissimilar from Aristotle's contention that virtue is a matter of practice. There are quite simple ways of doing this, including taking some time at the end of the day to mentally go through the good things you have accomplished (it worked for Julius Caesar apparently!) or to pay "gratitude visits," that is, to take the time to thank (not necessarily in person, given our electronic age) people who have influenced you positively or have otherwise done good things for you. "Mindfulness" is another way to find meaning in life, and it can be practiced in a variety of ways, from meditation practices to the exercise of paying attention to what you do and reflecting on why you do it (in other words, doing philosophy). All of this, by the way, will result in living not only a happier life (as if that were already not enough!) but also a longer one: researchers have found that people who train themselves to engage with positive emotions live about ten years longer than those who dwell on negative ones. That's the same order of difference that we find in life expectancy between smokers and nonsmokers (which does not at all mean that I am advising you to take up smoking as long as you think positively about it!).

Sociologists have gotten into "happiness research" too, of late, and one wonders exactly why it took the social sciences so long to figure out that it might be interesting—and useful—to see what makes people's lives better and more meaningful. Some of the results are not at all surprising, but others will give you something unexpected to ponder. The first thing that may come as a surprise to many Americans, though it has actually been known for a while, is that there is little relationship between wealth and happiness, with some important caveats. At a societal level, for instance, the gross domestic product (GDP) of the United States steadily increased from 1978 to 2008 (to be precise, it went from $2.3 trillion to $14.4 trillion), and yet measures of self-reported happiness stayed about the same or even declined during the same period. That's why in recent years the United Nations has started to produce statistics on what it calls the "human development index," a more comprehensive measure that includes data not just on GDP but also on health and education.

This is not to say that "money can't buy you happiness," however, just that (a certain amount of) money by itself is necessary but not sufficient for happiness. Research worldwide has confirmed that we do need a minimum level of income and amenities (such as a house and basic health care) in order to be happy, but that beyond that minimum the amount of our wealth quickly becomes a poor statistical predictor of our happiness. (If you guessed that I'm now going to say, "Aristotle said so," you are correct: the philosopher made clear that, contrary to what the Stoics were saying around his time, one does need some basic comforts in life in order to pursue eudaimonia, so that happiness is, to some extent, also a matter of luck and circumstance.)

Interestingly, researchers in the United States have been able to quantify the effect of extra income on self-reported happiness, and the results are amusing, to say the least. It turns out, for instance, that every extra $1,000 corresponds roughly to an increase of 0.002 on a social science index of happiness. To put this in context, it means that if you make an extra $100,000, your happiness will increase by about the same quantity that separates married (happier) from unmarried (unhappier) people, or employed (happier) from unemployed (unhappier) ones. I doubt, however, that the increase is linear, or indeed can even be sustained—I wouldn't expect someone who makes, say, over a million a year to be that much happier if he manages to add another $100,000 to his bank account. And of course one has to discount these findings to some extent because wealthier people may feel somewhat obliged to report being happier, especially if being wealthy is an important part of their ethos and self-image (which I would bet it often is).

Recent surveys on happiness in the United States are illuminating in terms of what researchers call the "statistical structure of subjective well-being," that is, the factors that seem to influence our happiness. Take the following examples with the mandatory grain of salt that correlation does not necessarily imply causation (though the two are highly correlated): women tend to be happier than men; predictably, wealthier, healthier, and more educated people are happier; married people are happier than unmarried ones; and whites are happier than any other ethnic group (again, in the United States). Exercising and eating fruit is associated with happiness, while being fat has a negative relation with subjective well-being. Oh, and having children in your household, though undoubtedly adding meaning to your life, has a surprisingly negative effect on your happiness. Moreover, remember that nonsense about women being from Venus and men from Mars? Well, actual research shows that men's and women's happiness seems to be affected pretty much by the same factors, in pretty much the same way. It appears that we are from the same planet after all.

What happens when we compare subjective well-being across the world? It turns out that the happiest nations are Ireland, Switzerland, Mexico, the United States, Great Britain, New Zealand, Denmark, Sweden, Finland, Norway, Luxembourg, and the Netherlands. (Notice the disproportionate number of European countries, particularly Scandinavian ones.) The unhappiest places in the world include Russia, Bulgaria, Latvia, Croatia, Hungary, and Macedonia (all in, essentially, eastern Europe). What is most interesting, however, is to find out what the happiest places have in common in terms of indicators of individual happiness. Again, the list is not entirely surprising, but here it is: low unemployment and inflation, low inequality, strong welfare states, high public spending, low pollution, high levels of democratic participation, and strong networks of friends. In other words, people are happiest in precisely the countries that best approximate John Rawls's just state (Chapter 15).

One of the results from these studies that I find most intriguing is that age and life satisfaction are related in a complex manner best described by a U-shaped function. In the United States, you are likely to be most unhappy around age forty, though in Europe the same curve bottoms out around age fifty-four. Hence the legendary "midlife crisis" that so many people go through (particularly men, who, remember, tend to be less happy than women—other things being equal). What is stunning here is that life satisfaction then keeps climbing all the way into one's late eighties, if one is lucky enough to live that long. I realize that you might be getting a bit tired of hearing about Aristotle, but he did say that eudaimonia is a lifelong project, the outcome of which can be determined only after one dies.

Which brings me to the last topic of our exploration of meaning and life: the relationship between age and wisdom. Wisdom, these days, is a rather old-fashioned term, though of course achieving it was the whole point of both Western and Eastern ancient philosophies (whether people called it "eudaimonia" or "enlightenment"). Wisdom cannot be equated with factual knowledge, and certainly not with technical knowledge in any particular area of application. Indeed, many older people, especially in these days of fast scientific and technological advancement, commonly know much less about a number of domains than bright young kids do. But nobody would think it likely that a seventeen-year-old computer whiz is wiser than someone who has lived on this planet for a number of decades. That's because wisdom has to do with experiential knowledge of human social situations, a type of knowledge that can only come, well, with experience. And a major running theme throughout this book has been that coupling that experience with philosophical reflection on its meaning, as well as with the best information that modern science is capable of providing, amounts to the most powerful way to navigate our existence in an intelligent fashion.

Columbia University's Vivian Clayton has done a lot of research on wisdom, beginning back in the 1970s, when she was perhaps the first researcher to suggest that the topic was amenable to scientific investigation. According to her, wisdom can be thought of as having three underlying components: the acquisition of knowledge (a cognitive function), the analysis of that knowledge (a reflective function), and the filtering of that knowledge through the emotions (an affective function). In other words, in order to be wise one needs to know things, to think about them, and to calibrate that knowledge through one's emotional responses.

Subsequent research has shown that older and wiser (the two are not automatically correlated!) people tend to learn from their negative experiences and that they are capable of distinguishing situations when it makes sense to take some kind of action from situations that simply need to be accepted because there are no viable alternatives. They focus better than younger and less wise individuals on goals that are emotionally meaningful, and neurologically speaking they exercise control over their amygdala (the brain center of emotional response), using their prefrontal cortex (the brain's executive function); they end up spending more time on positive emotions and avoiding negative ones. Philosopher and psychologist William James was probably getting at something like this when he said, "The art of being wise is the art of knowing what to overlook."

SO WE COME TO THE END OF A JOURNEY DURING WHICH WE have used the best of what science and philosophy have been able to tell us so far about how the world works and how to find our place in it. Along the way we have seen that there are probably no gods, and that even if there are, they can't tell us how to be moral, nor can they give meaning to our lives. We have seen that love and friendship are biologically grounded, crucial for our existence, and a constant source of philosophical challenges. We have discussed the evolutionary basis and the neurobiological underpinnings of morality, neither of which excuses us from thinking hard about what the right things to do are for us as individuals, as well as what kind of society we want and why. We have learned that willpower, and indeed our very ability to take conscious charge of our lives—the awareness on which philosophical reflection is founded—are much more limited than previously thought. And we have seen that even science itself is subject to constraints about knowledge and certainty.

The latter two points in particular raise a potentially serious objection to the whole idea of sci-phi. On the one hand, we have the undeniable fact that scientific knowledge is always provisional, which implies that some, perhaps even most, of the specific scientific claims you have read about in this book may be out of date in a year, or a decade. On the other hand, we have seen that philosophers disagree on what to think about the sort of questions we have been considering, which means that there is no established philosophical truth we can use as a bedrock for our reflections on life's meaning. Compare this double source of uncertainty with the apparent stability of religious or mystical teachings over a span of centuries or millennia. When I was living in Knoxville, Tennessee, a local preacher was very upset about my writings and lectures about evolution and wrote an angry letter to the editor of the local paper. In the letter, the frustrated preacher wondered out loud about why some people prefer the constant uncertainties of a science that keeps changing to the bedrock certainties contained in the Bible. Good question, and we need to address it.

To begin with, of course, the Bible—like any other religious text—needs to be interpreted, and the interpretation demonstrably changes with the cultural milieu. Contrary to the stated belief of some religious fundamentalists, there is no such thing as a literal reading of scriptures, as shown by the perennial disagreement among religious sects over what particular scriptural passages actually mean. Moreover, there are plenty of teachings in the Bible—for instance, the injunction to kill children who disrespect their parents (found, among other places, in both Exodus 21:17 and Matthew 15:4)—that are in plain sight for anyone to read about, but that most people simply ignore as a vestige of a more barbarous time.

We also need to consider that there is no rational reason to accept the authority of any religious text at all. Not only can gods not possibly be the ultimate source of morality; not only is there very good reason for any rational person to doubt the existence of the supernatural to begin with; but more importantly, any such authority would have to be mediated by human agents (priests, preachers, rabbis, imams, gurus, and the like), and such mediation seems to be hopelessly subjective and open to far more doubts than the reasonings of philosophers or the tentative conclusions of scientists. At least one can argue with the former on the basis of logic and question the latter on the grounds of empirical evidence.

But there is a more fundamental reason why sci-phi's tentativeness is not fatal: far from being a problem with that approach, it is in fact its primary virtue. We need to wrap our minds around the fact that as human beings we are inherently limited in our ability to reason and to discover things about the world. These limitations do not give us a license to arbitrarily "go beyond" reason and evidence into religion or mysticism. On the contrary, they are reminders that nobody

Conclusion: Human Nature and the Meaning of Life

has final answers and that the quest is open to all people who are willing to use their brain intelligently. Our limitations also give us a reason to cut ourselves a bit of slack for not getting life exactly right, for failing here and there, as humans are bound to do. This is why the eudaimonic life is always an imperfect and incomplete project, all the way until the moment of our death. But it is by far the most important of our projects, and one for which sci-phi is far better equipped to help us along the way than simple common sense, political ideology, or religious mysticism. We are social and (somewhat) rational animals, and we can reflect on how to employ our rationality to improve our lives and our societies. Seems like the meaningful thing to do.

ACKNOWLEDGMENTS

In writing a book, one is influenced and helped, directly or indirectly, by a large number of people. I wish in particular to recognize a few here. To begin with, my former postdoctoral associate at Stony Brook University, Oliver Bossdorf, for actually having suggested the original idea for this book, based on an informal talk built around Monty Python songs having to do with philosophical themes. Also many thanks to my collaborators at the Rationally Speaking blog (rationallyspeaking.org) and podcast (rationallyspeakingpodcast.org): Benny Pollak, the podcast's producer, and Julia Galef, my cohost, Phil Pollack, the blog's editor, and my cowriters there: Michael De Dora, Leonard Finkleman, Tunç Iyriboz, Greg Linster, and Ian Pollock. All of these friends have challenged my ideas in the kind of friendly and productive way that inspired the motto of the blog: "Truth springs from argument amongst friends" (David Hume). Special thanks also to Judy Heiblum, my agent, and T. J. Kelleher, my editor, for having worked so patiently with me on this project. My love of science was nurtured by my adoptive grandfather, Tino Soraci, who is no longer with us, and my appreciation for philosophy was first sparked by my high school teacher Enrica Chiaramonte. Early influences in one's life often go a long way indeed, as Aristotle surely appreciated.

DIGGING DEEPER

Chapter 1: Sci-Phi and the Meaning of Life

A Treatise of Human Nature [1740] by David Hume. Dover, 2003.

Rethinking Thin: The New Science of Weight Loss—and the Myths and Realities of Dieting by Gina Kolata. Picador, 2008.

Nonsense on Stilts: How to Tell Science from Bunk by Massimo Pigliucci. University of Chicago Press, 2010.

Chapter 2: Trolley Dilemmas and How We Make Moral Decisions

Logicomix by Apostolos Doxiadis and Christos Papadimitriou. Bloomsbury, 2009.

"Morality and Evolutionary Biology" by William Fitzpatrick. *Stanford Encyclopedia of Philosophy,* December 19, 2008, http://plato.stanford.edu/entries/morality-biology.

"From Neural 'Is' to Moral 'Ought': What Are the Moral Implications of Neuroscientific Moral Psychology?" by Joshua Greene. *Nature Reviews Neuroscience* 4(2003): 847–850.

"The Emotional Dog and Its Rational Tail: A Social Intuitionist Approach to Moral Judgment" by Jonathan Haidt. *Psychological Review* 108(2001): 814–834.

"Damage to the Prefrontal Cortex Increases Utilitarian Moral Judgements" by Michael Koenigs, Liane Young, Ralph Adolphs, Daniel Tranel, Fiery Cushman, Marc Hauser, and Antonio Damasio. *Nature* 446(2007): 908–911.

"Is Ethics a Science?" by Massimo Pigliucci. *Philosophy Now* 55(2006): 25.

"On the Relationship Between Science and Ethics" by Massimo Pigliucci. *Zygon* 38(2003): 871–894.

"The Trolley Problem" by Judith J. Thomson. *Yale Law Journal* 94(May 1985): 1395–1415.

Chapter 3: Your Brain on Morality

"Neuroscience and Ethics: Intersections" by Antonio Damasio. *American Journal of Bioethics* 7(2007): 3–7.

"Can Genes and Brain Abnormalities Create Killers?" by Neal Conan, with Barbara Bradley Hagerty. National Public Radio, *Talk of the Nation*, July 6, 2010.

"Neuroanatomical Background to Understanding the Brain of the Young Psychopath" by James H. Fallon. *Ohio State Journal of Criminal Law* 3(2006): 341–367.

"The Cognitive Neuroscience of Moral Judgment" by Joshua Greene. In *The Cognitive Neurosciences IV*. MIT Press, 2009.

"The Right and the Good: Distributive Justice and Neural Encoding of Equity and Efficiency" by Ming Hsu, Cédric Anen, and Steven R. Quartz. *Science* 320(2008): 1092–1095.

"The Role of Emotion in Moral Psychology" by Bryce Huebner, Susan Dwyer, and Marc Hauser. *Trends in Cognitive Science* 13(2009): 1–6.

"Moral Judgment and the Brain: A Functional Approach to the Question of Emotion and Cognition in Moral Judgment Integrating Psychology, Neuroscience, and Evolutionary Biology" by

Kristin Prehn and Hauke Heekeren. In *The Moral Brain*, edited by Jan Verplaetse et al., pp. 129–154. Springer, 2009.

Chapter 4: The Evolution of Morality

"The Evolution of Cooperation" by Robert Axelrod and William D. Hamilton. *Science* 211(1981): 1390–1396.

The Selfish Gene by Richard Dawkins. Oxford University Press, 1976.

"'Any Animal Whatever': Darwinian Building Blocks of Morality in Monkeys and Apes" by Jessica C. Flack and Frans B. M. de Waal. *Journal of Consciousness Studies* 7(2000): 1–29.

Leviathan [1651] by Thomas Hobbes. http://www.gutenberg.org/ebooks/3207.

Evolution and Ethics [1894] by Thomas Henry Huxley. http://www.gutenberg.org/ebooks/2940.

"Evolution of Indirect Reciprocity" by Martin Nowak and Karl Sigmund. *Nature* 437(2005): 1291–1298.

The Expanding Circle: Ethics and Sociobiology by Peter Singer. Clarendon, 1981.

The Folly of Fools: The Logic of Deceit and Self-Deception in Human Life by Robert Trivers. Basic Books, 2011.

Primates and Philosophers: How Morality Evolved by Frans B. M. de Waal. Princeton University Press, 2006.

Adaptation and Natural Selection by George Williams. Princeton University Press, 1996.

Chapter 5: A Handy-Dandy Menu for Building Your Own Moral Theory

"Deontological Ethics" by Larry Alexander and Michael Moore. *Stanford Encyclopedia of Philosophy*, November 21, 2007, http://plato.stanford.edu/entries/ethics-deontological.

"Well-Being" by Roger Crisp. *Stanford Encyclopedia of Philosophy*, December 9, 2009, http://plato.stanford.edu/entries/well-being.

The Moral Landscape: How Science Can Determine Human Values by Sam Harris. Free Press, 2011.

"Virtue Ethics" by Rosalind Hursthouse. *Stanford Encyclopedia of Philosophy*, July 18, 2007, http://plato.stanford.edu/entries/ethics-virtue.

The Basic Writings of John Stuart Mill: On Liberty, the Subjection of Women, and Utilitarianism. CreateSpace, 2009.

"Moral Luck" [1979] by Thomas Nagel. http://philosophyfaculty.ucsd.edu/faculty/rarneson/Courses/NAGELMoralLuck.pdf.

Kant: A Very Short Introduction by Roger Scruton. Oxford University Press, 2001.

"The Triviality of the Debate over 'Is-Ought' and the Definition of 'Moral'" by Peter Singer. *American Philosophical Quarterly* 10(January 1973), http://www.utilitarian.net/singer/by/197301—.htm.

"Consequentialism" by Walter Sinnott-Armstrong. *Stanford Encyclopedia of Philosophy*, September 27, 2011, http://plato.stanford.edu/entries/consequentialism.

Virtue Ethics: An Introduction by Richard Taylor. Prometheus Books, 2002.

Chapter 6: The Not So Rational Animal

"It Feels Like We're Thinking: The Rationalizing Voter and Electoral Democracy" by Christopher H. Achen and Larry M. Bartels. Paper presented to the annual meeting of the American Political Science Association, 2006.

"Delusion" by Lisa Bortolotti. *Stanford Encyclopedia of Philosophy*, September 16, 2009, http://plato.stanford.edu/entries/delusion (for the account of the woman with Cotard's syndrome).

"Decision Making: Rational or Hedonic?" by Michel Cabanac and Marie-Claude Bonniot-Cabanac. *Behavioral and Brain Functions* 3(2007).

"Frames, Biases, and Rational Decision-Making in the Human Brain" by Benedetto de Martino, Dharshan Kumaran, Ben Seymour, and Raymond J. Dolan. *Science* 313(2006): 684–687.

"The Split Brain Revisited" by Michael Gazzaniga. *Scientific American* 287(2003): 26–31.

Phantoms in the Brain by V. S. Ramachandran. HarperCollins, 1999.

"Spontaneous Confabulation and the Adaptation of Thought to Ongoing Reality" by Armin Schnider. *Nature Reviews Neuroscience* 4(2003): 662–671.

Being Wrong: Adventures in the Margin of Error by Kathryn Schulz. Ecco, 2010.

Mistakes Were Made (but Not by Me): Why We Justify Foolish Beliefs, Bad Decisions, and Hurtful Acts by Carol Tavris and Elliot Aronson. Mariner Books, 2008.

Chapter 7: Intuition Versus Rationality, and How to Become Really Good at What You Do

"Overcoming Intuition: Metacognitive Difficulty Activates Analytic Reasoning" by Adam L. Alter, Daniel M. Oppenheimer, Nicholas Epley, and Rebecca N. Eyre. *Journal of Experimental Psychology: General* 136(2007): 569–576.

"Which Should You Use, Intuition or Logic? Cultural Differences in Injunctive Norms About Reasoning" by Emma E. Buchtel and Ara Norenzayan. *Asian Journal of Social Psychology* 11(2008): 264–273.

"The Influence of Experience and Deliberate Practice on the Development of Superior Expert Performance" by K. Anders Ericsson. In *The Cambridge Handbook of Expertise and Expert Performance*, edited by K. Anders Ericsson et al. Cambridge University Press, 2006.

Delusions of Gender by Cordelia Fine. W. W. Norton, 2011.

"Comparing Expert and Novice Understanding of a Complex System from the Perspective of Structures, Behaviors, and Functions" by Cindy E. Hmelo-Silver and Merav Green Pfeffer. *Cognitive Science* 28(2004): 127–138.

"Intuition: A Fundamental Bridging Construct in the Behavioural Sciences" by Gerard P. Hodgkinson, Janice Langan-Fox, and Eugene Sadler-Smith. *British Journal of Psychology* 99(2008): 1–27.

"Intuition and the Correspondence Between Implicit and Explicit Self-Esteem" by Christian H. Jordan, Mervyn Whitfield, and Virgil Zeigler-Hill. *Journal of Personality and Social Psychology* 93(2007): 1067–1079.

Brain Storm: The Flaws in the Science of Sex Differences by Rebecca M. Jordan-Young. Harvard University Press, 2011.

"The Expert Mind" by Philip E. Ross. *Scientific American*, July 24, 2006.

The Genius in All of Us: Why Everything You've Been Told About Genetics, Talent, and IQ Is Wrong by David Shenk. Doubleday, 2010.

"Intuition: Myth or a Decision-Making Tool?" by Marta Sinclair and Neal M. Ashkanasy. *Management Learning* 36(2005): 353–370.

Chapter 8: The Limits of Science

What Is This Thing Called Science? by Alan Chalmers. University of Queensland Press, 1999.

Scientific Perspectivism by Ronald Giere. University of Chicago Press, 2010.

"Science and Pseudo-Science" by Sven Ove Hansson. *Stanford Encyclopedia of Philosophy,* September 3, 2008, http://plato.stanford.edu/entries/pseudo-science.

The Structure of Scientific Revolutions by Thomas Kuhn. University of Chicago Press, 1962.

Understanding Philosophy of Science by James Ladyman. Routledge, 2002.

Philosophy of Science: A Very Short Introduction by Samir Okasha. Oxford University Press, 2002.

Conjectures and Refutations: The Growth of Scientific Knowledge by Karl Popper. Routledge, 1963.

"The Problem of Induction" by John Vickers. *Stanford Encyclopedia of Philosophy,* June 21, 2010, http://plato.stanford.edu/entries/induction-problem.

Chapter 9: The (Limited) Power of the Will

Elbow Room: The Varieties of Free Will Worth Wanting by Daniel Dennett. MIT Press, 1984.

"The Sources of Human Volition" by Patrick Haggard. *Science* 324(2009): 731–733.

"Is Free Will an Illusion?" by Martin Heisenberg. *Nature* 459(2009): 164–165.

"Anterior Prefrontal Function and the Limits of Human Decision Making" by Etienne Koechlin and Alexandre Hyafil. *Science* 318(2007): 594–598.

"Unconscious Cerebral Initiative and the Role of Conscious Will in Voluntary Action" by Benjamin Libet. *Behavioral and Brain Sciences* 8(1985): 529–566.

"Religion, Self-Control, and Self-Regulation: Associations, Explanations, and Implications" by Michael McCullough and Brian Willoughby. *Psychological Bulletin* 135(2009): 69–93.

"Personality Traits and Cancer Risk and Survival Based on Finnish and Swedish Registry Data" by Naoki Nakaya et al. *American Journal of Epidemiology* 172(2010): 377–385.

"Free Will" by Timothy O'Connor. *Stanford Encyclopedia of Philosophy*, October 29, 2010, http://plato.stanford.edu/entries/freewill.

"Free Choice Activates a Decision Circuit Between Frontal and Parietal Cortex" by Bijan Pesaran, Matthew J. Nelson, and Richard A. Andersen. *Nature* 453(2008): 406–410.

"How Does Neuroscience Affect Our Conception of Volition?" by Adina L. Roskies. *Annual Review of Neuroscience* 33(2010): 109–130.

Chapter 10: Who's in Charge Anyway? The Zombie Inside You

"Living on Impulse" by Benedict Carey. *New York Times*, April 4, 2006.

"Who's Minding the Mind?" by Benedict Carey. *New York Times*, July 31, 2007.

"In Bad Taste: Evidence for the Oral Origins of Moral Disgust" by H. A. Chapman, D. A. Kim, J. M. Susskind, and A. K. Anderson. *Science* 323(2009): 1222–1226.

"The Currency of Guessing" by Paul Cisek. *Nature* 447(2007): 1061–1062.

"Hume's Moral Philosophy" by Rachel Cohon. *Stanford Encyclopedia of Philosophy*, October 29, 2004, http://plato.stanford.edu/entries/hume-moral.

"Dispositional Impulsivity in Normal and Abnormal Samples" by Janine D. Flory, Philip D. Harvey, Vivian Mitropoulou, Antonia S. New, Jeremy M. Silverman, Larry J. Siever, and Stephen B. Manuck. *Journal of Psychiatric Research* 40(2006): 438–447.

Beyond the Pleasure Principle [1920] by Sigmund Freud. http://www.bartleby.com/276/.

A Treatise on Human Nature [1739–1740] by David Hume. http://www.gutenberg.org/ebooks/4705.

"Time of Conscious Intention to Act in Relation to Onset of Cerebral Activity (Readiness-Potential): The Unconscious Initiation of a Freely Voluntary Act" by Benjamin Libet, Curtis A. Gleason, Elwood W. Wright, and Dennis K. Pearl. *Brain* 106(1983): 623–642.

"Ancient Theories of Soul" by Hendrik Lorenz. *Stanford Encyclopedia of Philosophy*, April 22, 2009, http://plato.stanford.edu/entries/ancient-soul.

"How the Brain Translates Money into Force: A Neuroimaging Study of Subliminal Motivation" by Mathias Pessiglione, Liane Schmidt, Bogdan Draganski, Raffael Kalisch, Hakwan Lau, Ray J. Dolan, and Chris D. Frith. *Science* 316(May 2007): 904–906.

The Republic by Plato, edited by R. W. Sterling and W. C. Scott. W. W. Norton, 1996.

"From Oral to Moral" by Paul Rozin, Jonathan Haidt, and Katrina Fincher. *Science* 323(2009): 1179–1180.

"Probabilistic Reasoning by Neurons" by T. Yang and M. N. Shadlen. *Nature* 447(2007): 1075–1080.

Chapter 11: The Hormones of Love

Why We Love: The Nature and Chemistry of Romantic Love by Helen E. Fisher. H. Holt, 2004.

"Lust, Romance, Attachment: Do the Side Effects of Serotonin-Enhancing Antidepressants Jeopardize Romantic Love, Marriage, and Fertility?" by Helen E. Fisher and J. Anderson Thomson Jr. In *Evolutionary Cognitive Neuroscience*, edited by Steven M. Platek, Julian Paul Keenan, and Todd K. Shackelford. MIT Press, 2006.

"Love" by Bennett Helm. *Stanford Encyclopedia of Philosophy*, July 9, 2009, http://plato.stanford.edu/entries/love.

"Love: You Have Four Minutes to Choose Your Perfect Mate" by Matt Kaplan. *Nature* 451(2008): 760–762.

"The Myth, the Math, the Sex" by Gina Kolata. *New York Times*, August 12, 2007.

"A Love Vaccine?" by John Tierney. *New York Times*, January 12, 2009.

"MHC-Dependent Mate Preferences in Humans" by Claus Wedekind, Thomas Seebeck, Florence Bettens, and Alexander J. Paepke. *Proceedings of the Royal Society: Biological Science* 260(1995): 245–249.

"Love: Neuroscience Reveals All" by Larry J. Young. *Nature* 457(2009): 148.

Chapter 12: Friendship and the Meaning of Life

"Friendship as a Health Factor" by Jennifer Couzin. *Science* 323(2009): 454–457.

"An Economic Model of Friendship: Homophily, Minorities, and Segregation" by Sergio Currarini, Matthew O. Jackson, and Paolo Pin. *Econometrica* 77(2009): 1003.

"'I Am So Happy 'Cause Today I Found My Friend': Friendship and Personality as Predictors of Happiness" by Meliksah Demir and Lesley A. Weitekamp. *Journal of Happiness Studies* 8(2007): 181–211.

"Inferring Friendship Network Structure by Using Mobile Phone Data" by Nathan Eagle, Alex "Sandy" Pentland, and David Lazer. *Proceedings of the National Academy of Science–USA* 106(2009): 15274–15278.

"Community Structure and Ethnic Preferences in School Friendship" by Marta C. Gonzalez, Hans J. Herrmann, Janos Kertesz, and Tamás Vicsek. *Physica* 379(2007): 307–316.

"Cooperative Behavior Cascades in Human Social Networks" by James H. Fowler and Nicholas A. Christakis. *Proceedings of the National Academy of Sciences* 107(2010): 5334–5338.

"Friendship Versus Business in Marketing Relationships" by Kent Grayson. *Journal of Marketing* 71(2007): 121–139.

"Friendship" by Bennett Helm. *Stanford Encyclopedia of Philosophy*, July 9, 2009, http://plato.stanford.edu/entries/friendship.

Obedience to Authority: An Experimental View by Stanley Milgram. Harper & Row, 1974.

"Virtual Friendship and the New Narcissism" by Christine Rosen. *The New Atlantis* (Summer 2007): 15–31.

Six Degrees: The Science of a Connected Age by Duncan J. Watts. W. W. Norton, 2003.

Chapter 13: Right, Left, Up, Down: On Politics

"Partisans Without Constraint: Political Polarization and Trends in American Public Opinion" by Delia Baldassarri and Andrew Gelman. *American Journal of Sociology* 114(2008): 408–544.

"Political Belief Networks: Socio-Cognitive Heterogeneity in American Public Opinion" by Delia Baldassarri and Amir Goldberg. Political Networks Paper Archive Working Papers, August 1, 2010, http://opensiuc.lib.siu.edu/cgi/viewcontent.cgi?article=1050&context=pn_wp.

"Political Interest, Cognitive Ability, and Personality: Determinants of Voter Turnout in Britain" by Kevin Denny and Orla Doyle. Geary Institute Discussion Paper Series 2005/07, http://www.ucd.ie/geary/publications/2005/GearyWp200507.pdf.

"Political Ideology: Its Structure, Functions, and Elective Affinities" by John T. Jost, Christopher Federico, and Jaime Napier. *Annual Review of Psychology* 60(2009): 307–357.

"Political Orientations Are Correlated with Brain Structure in Young Adults" by Ryota Kanai, Tom Feilden, Colin Firth, and Geraint Rees. *Current Biology* 21(April 26, 2011): 1–4.

"Sticking with Your Vote: Cognitive Dissonance and Political Attitudes" by Sendhil Mullainathan and Ebonya Washington. *American Economic Journal: Applied Economics* 1(2009): 86–111.

"Political Attitudes Vary with Physiological Traits" by Douglas Oxley, Kevin Smith, John Alford, Matthew Hibbing, Jennifer Miller, Mario Scalora, Peter Hatemi, and John R. Hibbing. *Science* 321(2008): 1667–1670.

"'There Must Be a Reason': Osama, Saddam, and Inferred Justification" by Monica Prasad, Andrew Perrin, Kieran Bezila, Steve Hoffman, Kate Kindleberger, Kim Manturuk, and Ashleigh Powers. *Sociological Inquiry* 79(2009): 142–163.

"More Than Weighting Cognitive Importance: A Dual-Process Model of Issue Framing Effects" by Rune Slothuus. *Political Psychology* 29(2008): 1–28.

Chapter 14: Our Innate Sense of Fairness

"Serotonin Modulates Behavioral Reactions to Unfairness" by Molly J. Crockett, Luke Clark, Golnaz Tabibnia, Matthew D. Lieberman, and Trevor W. Robbins. *Science* 320(2008): 1739.

"Reflective Equilibrium" by Norman Daniels. *Stanford Encyclopedia of Philosophy*, January 12, 2011, http://plato.stanford.edu/entries/reflective-equilibrium.

"Egalitarianism in Young Children" by Ernst Fehr, Helen Bernhard, and Bettina Rockenbach. *Nature* 454(2008): 1079–1083.

"The Right and the Good: Distributive Justice and Neural Encoding of Equity and Efficiency" by Ming Hsu, Cédric Anen, and Steven R. Quartz. *Science* 320(May 23, 2008): 1092–1095.

"Share and Share Alike" by Michael Tomasello and Felix Warneken. *Nature* 454(2008): 1057–1058.

Chapter 15: On Justice

"What Are Human Rights? Four Schools of Thought" by Marie-Bénédicte Dembour. *Human Rights Quarterly* 32(2010): 1–20.

"The Competitive Advantage of Sanctioning Institutions" by Ozgur Gurerk, Bernd Irlenbusch, and Bettina Rockenbach. *Science* 312(2006): 108–111.

"The Free Rider Problem" by Russell Hardin. *Stanford Encyclopedia of Philosophy*, May 21, 2003, http://plato.stanford.edu/entries/free-rider.

"Are Our Minds Fundamentally Egalitarian? Adaptive Bases of Different Socio-Cultural Models About Distributive Justice" by Tatsuya Kameda, Masanori Takezawa, Yohsuke Ohtsubo, and Reid Hastie. In *Evolution, Culture, and the Human Mind*, edited by Mark Schaller, Steven J. Heine, Ara Norenzayan, Toshio Yamagishi, and Tatsuya Kameda. Lawrence Erlbaum Associates, 2009.

A Theory of Justice by John Rawls. Harvard University Press, 1971.

Justice: What's the Right Thing to Do? by Michael Sandel. Farrar, Straus and Giroux, 2010.

"John Rawls" by Leif Wenar. *Stanford Encyclopedia of Philosophy*, March 25, 2008, http://plato.stanford.edu/entries/rawls.

Chapter 16: Your Brain on God

"Stimulating Illusory Own-Body Perceptions" by Olaf Blanke, Stéphanie Ortigue, Theodor Landis, and Margitta Seeck. *Nature* 419(2002): 269–270.

Descartes' Baby: How the Science of Child Development Explains What Makes Us Human by Paul Bloom. Basic Books, 2004.

"Born Believers: How Your Brain Creates God" by Michael Brooks. *New Scientist*, February 4, 2009.

"From Haunted Brain to Haunted Science: A Cognitive Neuroscience View of Paranormal and Pseudoscientific Thought" by

Peter Brugger. In *Hauntings and Poltergeists: Multidisciplinary Perspectives*, edited by James Houran and Rense Lange. McFarland & Company, 2001.

"Keep Your Fingers Crossed! How Superstition Improves Performance" by Lysann Damisch, Barbara Stoberock, and Thomas Mussweiler. *Psychological Science*, published online, May 28, 2010, doi:10.1177/0956797610372631.

"Are Spiritual Encounters All in Your Head?" by Barbara Bradley Hagerty. National Public Radio, *All Things Considered*, May 19, 2009, http://www.npr.org/templates/story/story.php?storyId=104291534 (on the connection between epilepsy and religious experiences).

"Intuitions About Origins: Purpose and Intelligent Design in Children's Reasoning About Nature" by Deborah Kelemen and Cara DiYanni. *Journal of Cognition and Development* 6(2005): 3–31.

Neuropsychological Bases of God Beliefs by Michael Persinger. Praeger, 1987.

"Paranormal Beliefs Linked to Brain Chemistry" by Helen Philips. NewScientist.com, July 27, 2002, http://www.newscientist.com/article/dn2589-paranormal-beliefs-linked-to-brain-chemistry.html.

"Lacking Control Increases Illusory Pattern Perception" by Jennifer A. Whitson and Adam D. Galinsky. *Science* 322(2008): 115–117.

Chapter 17: The Evolution of Religion

Gods We Trust: The Evolutionary Landscape of Religion by Scott Atran. Oxford University Press, 2002.

Religion Explained: The Evolutionary Origins of Religious Thought by Pascal Boyer. Basic Books, 2001.

Breaking the Spell: Religion as a Natural Phenomenon by Daniel Dennett. Penguin, 2007.

The Natural History of Religion [1757] by David Hume. EZReads Publications, 2010.

"The Origin and Evolution of Religious Prosociality" by Ara Norenzayan and Azim F. Shariff. *Science* 322(2008): 58–62.

"The Origins of Religion: Evolved Adaptation or By-Product?" by Ilkka Pyysiainen and Marc Hauser. *Trends in Cognitive Science* 14(2010): 104–109.

"'Superstition' in the Pigeon" by Burrhus F. Skinner. *Journal of Experimental Psychology* 38(1948): 168–172.

God's Brain by Lionel Tiger and Michael McGuire. Prometheus, 2010.

Chapter 18: Euthyphro's Dilemma: Morality as a Human Problem

"Plato's Ethics: An Overview" by Dorothea Frede. *Stanford Encyclopedia of Philosophy*, May 29, 2009, http://plato.stanford.edu/entries/plato-ethics.

"Socrates" by Debra Nails. *Stanford Encyclopedia of Philosophy*, November 7, 2009, http://plato.stanford.edu/entries/socrates.

Euthyphro by Plato, translated by Benjamin Jowett, http://classics.mit.edu/Plato/euthyfro.html.

Conclusion: Human Nature and the Meaning of Life

"International Happiness" by David G. Blanchflower and Andrew J. Oswald. Working paper 16668. National Bureau of Economic Research, January 2011.

"Wisdom and Intelligence: The Nature and Function of Knowledge in the Later Years" by Vivian Clayton. *International Journal of Aging and Human Development* 15(1982): 315–321.

"Biology, Politics, and the Emerging Science of Human Nature" by James H. Fowler and Darren Schreiber. *Science* 322(2008): 912–914.

"Hume's Progressive View of Human Nature" by Michael Gill. *Hume Studies* 26(2000): 87–108.

"The Older-and-Wiser Hypothesis" by Stephen Hall. *New York Times,* May 6, 2007.

"The Other Strand" by Erika Hayden. *Nature* 457(2009): 776–779.

"Historical Evidence and Human Adaptations" by Jonathan Kaplan. *Philosophy of Science* 69(2002): S294–S304.

"Happiness 101" by D. T. Max. *New York Times,* January 7, 2007.

"Ten Politically Incorrect Truths About Human Nature" by Alan S. Miller and Satoshi Kanazawa. *Psychology Today,* July 1, 2007, http://www.psychologytoday.com/articles/200706/ten-politically-incorrect-truths-about-human-nature.

INDEX

akrasia (weakness of the will), 7, 71, 73, 131
altruism, 49–51
antidepressants (use of, effect on relationships), 170–171
Aquinas, Thomas, 269–270
Aristotle, 6, 71, 78, 89, 131, 160, 176–177, 187, 246, 273, 283
Aristotelian causes (of phenomena), 246–247

Bayes theorem, 196–197
Bentham, Jeremy, 15, 22, 68

Capgras syndrome, 82
categorical imperative (in ethics), 66–67
cognitive dissonance, 84, 195–196
   and beliefs about 9/11 and Saddam Hussein, 196–197
   defensive strategies to reduce, 198–199

cognitive load (or interference), 37, 97
confabulation (in psychology), 57, 84–85
Confucius, 101
Copernicus, Nicholas, 14
Cotard's syndrome, 81

Damasio, Antonio (neuroscientist), 23, 36
Darwin, Charles, 48
deductive reasoning, 111–112
deontology (rule-based ethics), 15, 23, 66–68
Dawkins, Richard (science writer), 46, 165, 250, 259
delusion (in cognitive science), 80–82
Dennett, Daniel (philosopher), 140
de Waal, Frans (primatologist), 48, 53, 187
disfluency (of cognitive efforts), 97–98

dual-process theory (of moral judgment), 35–38
  criticism of, 40–41

Epicurus, 8, 179, 254
eudaimonia (flourishing), 6–8, 72, 179, 279, 283, 287
Euthyphro's dilemma (about gods and morality), 266–268
  objections to, 269–272
evolutionary psychology (problems with), 274–275
expanding (moral) circle, 57–58, 65
expertise, 102–103
  about nothing, 107

Facebook, 181, 183, 246
fallacy of composition (and the free rider problem), 217
falsification (in philosophy of science), 115–116
Fitzpatrick, William (philosopher), 26–29
fMRI brain scans (problems with), 42–43
framing (in psychology and political science), 79–80, 88, 193–195
free rider problem (in ethics), 215–219
free will (types of), 142
  neurobiological basis of, 133–134

Libet experiment, 134–136, 146
  and determinism, 137–140
Freud, Sigmund, 148
  tri-partite theory of mind, 148
friendship
  effects of (spreading like disease), 175–176
  differences with love, 177
  Aristotle's types, 177–178
  as a mirror of our character, 179–180
  and ethical theories, 180–181
  and social networks, 181–183
fungibility (trading up)
  of love relations, 163
  of friends, 178–179

Galilei, Galileo, 9
Gazzaniga, Michael (neuroscientist), 83
God helmet, 232–233
Gödel, Kurt (logician), and incompleteness theorem, 28
Goodal, Jane (primatologist), 55
Google+, 181, 183
gossip (role in ethics), 52–53
Greene, Joshua (neuroscientist), 24, 35–36
Gyges's ring (mythology), 213–215

Haidt, Jonathan (psychologist), 25, 56–57

Haldane, J. B. S., 49
Hamilton, William Donald, 49
happiness
   set point for, 173
   and big five personality traits, 174
   and gender (no contribution of), 174
   and GDP, 280
   research on, 280–281
   and personal wealth, 281
   nations with higher and lower levels of, 282
Harris, Sam, 61–63
hedonic pressure, 86–87
hedonic treadmill, 279
Hippocrates, 232
Hobbes, Thomas, 45–46, 218
human nature
   according to David Hume, 275–276
   and the human genome project, 277–278
Hume, David, 8, 61, 112–114, 137, 150–151, 216, 245
Huxley, Thomas Henry, 46

inductive reasoning, 113–114
   and Russell's inductivist chicken/turkey, 113–114
   Hume's problem with, 113–114
intuition, 93–94
   brain areas involved with, 95
   and self-esteem, 95–96
   and business decision making, 98–99
   and women, 99–100
   inter-cultural differences, 100–102

James, William, 16, 93–94
justice as fairness, 210

Kant, Immanuel, 12, 66–68
Kepler, Johannes, 14
kin selection, 49
Kuhn, Thomas (philosopher), 117–118

language (evolution of), 249–251
likelihood ratio (as unconscious inferential tool), 144–145
Locke, John, 222
love
   three types recognized by the ancient Greeks, 159–160
   four concepts recognized by modern philosophers, 160–163
   three neurobiological phases of, 165–168

major histocompatibility complex (MHC) and love, 157–158
McCarthy, Jenny (celebrity), 111–112
memes (alleged units of cultural evolution), 249–250, 259–261

metaethics, 27, 60–65
mid-life crisis, 282–283
moral luck, 70, 72
mirror neurons (and empathy), 54
Mill, John Stuart, 15, 68, 216–217

Nagel, Thomas (philosopher), 70, 72
naturalistic fallacy (values from facts), 8–9, 61
natural selection, 26, 47, 49, 51, 55, 65, 250, 252–255, 261, 274
Neptune (planet), discovery of, 116
neuroethics, 32–33
Newton, Isaac, 9
Nietzsche, Friedrich, 14
"no miracles" argument (in philosophy of science), 120

obesity (weight problem), 1–4
"oral to moral" hypothesis (of emotive ethics), 152–153
out-of-body experiences, 233

paradigm shift (in philosophy of science), 117–118
Persinger, Michael (neuroscientist), 232–233
perspectivism (in philosophy of science), 122–123
pessimistic meta-induction (in philosophy of science), 121

philosophy (nature of), 12–15
Plato, 148, 213, 263–264
  theory of mind, 148–149
  on love, 157
political belief networks, 191–192
political preference
  and human physiology, 188
  and brain anatomy, 189
Popper, Karl (philosopher), 115–116
positive thinking (as pseudoscience), 130
post hoc ergo propter hoc (logical fallacy), 244
prisoner's dilemma (game theory), 216
proprioception (and mystical experiences), 233
pseudoscience (belief in), 110
priming (in psychology), 78, 96, 147
psychopathology (and ethical judgment), 31–33, 36–37

Radiolab (NPR show), 78–79
Ramachandran, V. S. (neuroscientist), 85
Rapaport, Anatol (psychologist), 50–51
Rawls, John (philosopher), 40, 205–206, 220–221
  "negative" thesis concerning justice as fairness, 224–225

"positive" thesis concerning justice as fairness, 225
and the two moral powers, 226
realism (and anti-realism) in philosophy of science, 118–121
reciprocal altruism, 50–51
reflecting equilibrium (a philosophical method), 205–206
narrow vs. broad application of, 207–208
and justice, 220–222
religion (evolution of), 253–261
as a by-product of human biology, 255–257
and cultural evolution, 258–261
repressed memories, 84–85

Sartre, Jean-Paul (philosopher), 138
science (nature of), 10–12
scientia (L., knowledge), 6, 62
sci-phi (science-philosophy), 2, 17, 62, 205, 220, 285–287
The Secret (book), 129
Skinner, B. F. (psychologist), 244
Singer, Peter (philosopher), 57–58, 69
six degrees of separation (Milgram experiment), 182–183

"smelly T-shirts" experiment (on love), 157–158
Socrates, 14, 157, 159, 215, 263–264
and Euthyphro (dialogue), 264–266, 268–269
split-brain patients, 82–84
string theory (in physics), 120
structured knowledge (and intuition, expertise), 104–106
superstition
and levels of dopamine, 235–236
and stress, 236–237
and unpredictability, 237
as placebo effect, 239–240
among pigeons (B. F. Skinner's experiment), 243–244
Swinburne, Richard (theologian), 270–271

talent (innate vs. acquired), 106–107
Ten Commandments (as a type of deontological ethics), 23
tit-for-tat (game theoretic strategy), 51, 64–65
tragedy of cognition (concerning superstition), 241
tragedy of the commons (in ethics and game theory), 216–217

Trivers, Robert (biologist), 50
trolley dilemmas, 21–23
   neurobiology of, 23–24
Twitter, 181, 183

ultimatum game (about fairness), 203–204
underdetermination of theory by data (in philosophy of science), 120
utilitarianism/consequentialism (ethical theory), 15, 22, 68–71

veil of ignorance (in Rawls's moral philosophy), 226–227
virtue ethics, 7–8, 71–73, 178, 276

Williams, George (biologist), 47
willpower (evidence for limited supply of), 130–133
Winfrey, Oprah, 129
wisdom, 283–284
Wittgenstein, Ludwig, 12